Digital Communications and Networks

Digital Communications and Networks

Edited by **Nelson Carter**

CWILLFORD PRESS

New York

Published by Willford Press,
118-35 Queens Blvd., Suite 400,
Forest Hills, NY 11375, USA
www.willfordpress.com

Digital Communications and Networks
Edited by Nelson Carter

International Standard Book Number: 978-1-68285-149-4 (Hardback)

Contents

Preface

Digital transmission is integrated deeply in our everyday life. Mobile networks, video conferences, wireless networks are all applications of digital communications. This book on digital communications and networks covers in detail some existent theories and innovative concepts, such as wireless communications, networking, information security, signal processing, synchronization and scheduling issues in mobile and ad hoc networks, etc. This book includes contributions of experts and scientists which will provide innovative insights into this field. Students, researchers, professionals and all associated with digital communication and allied disciplines will benefit alike from this book.

This book has been the outcome of endless efforts put in by authors and researchers on various issues and topics within the field. The book is a comprehensive collection of significant researches that are addressed in a variety of chapters. It will surely enhance the knowledge of the field among readers across the globe.

It gives us an immense pleasure to thank our researchers and authors for their efforts to submit their piece of writing before the deadlines. Finally in the end, I would like to thank my family and colleagues who have been a great source of inspiration and support.

Editor

1

An Intelligent Network Selection Strategy Based on MADM Methods in Heterogeneous Networks

Lahby Mohamed[1], Cherkaoui Leghris[2] and Adib Abdellah[3]

[1,2,3]Computer Science Department, LIM Lab
Faculty of Sciences and Technology of Mohammedia,
B.P. 146 Mohammedia, Morocco
Email: {mlahby, cleghris, adib_adbe}@yahoo.fr

ABSTRACT

Providing service continuity to the end users with best quality is a very important issue in the next generation wireless communications. With the evolution of the mobile devices towards a multimode architecture and the coexistence of multitude of radio access technologies (RAT's), the users are able to benefit simultaneously from these RAT's. However, the major issue in heterogeneous wireless communications is how to choose the most suitable access network for mobile's user which can be used as long as possible for communication.

To achieve this issue, this paper proposes an intelligent network selection strategy which combines two multi attribute decision making (MADM) methods such as analytic network process (ANP) and the technique for order preference by similarity to an ideal solution (TOPSIS) method. The ANP method is used to find the differentiate weights of available networks by considering each criterion and the TOPSIS method is applied to rank the alternatives. Our new strategy for network selection can dealing with the limitations of MADM methods which are the ranking abnormality and the ping-ponf effect.

KEYWORDS

Heterogeneous Wireless Network, Network Selection, Multi Attribute Decision Making, Ranking Abnormality, Ping Pong Effect.

1. INTRODUCTION

In recent years, the next generation wireless communications are growing rapidly and are integrating a multitude of radio access technologies (RAT's) such as wireless technologies (802.11a, 802.11b, 802.15, 802.16, etc.) and cellular networks (GPRS, UMTS, HSDPA, LTE, etc.) With the evolution of the mobile devices towards a multimode architecture and the coexistence of these heterogeneous RAT's the users are able to benefit simultaneously from these RAT's and they can also use various services offered by each type of access network.

However the most important issue in RAT's, is to provide ubiquitous access for the end users, under the principle "Always Best Connected" (ABC) [1], to achieve this issue a vertical handoff decision [2] is intended to determine whether a vertical handoff should be initiated, and to choose the most suitable network in terms of quality of service (QoS) for mobile users. The handover vertical process can be divided into three steps:

1) Handover initiation: it contains some preparation for handoff such as the measurement of received signal strength (RSS), QoS, security, battery level, etc.
2) Handover decision: it consists on choosing the most suitable network access among those available to perform a handover.
3) Handover execution: it consists on establishing the target access network by using mobile IP protocol (MIP).

The network selection problem is the most important key of the handover vertical decision. For that, our work focuses on the optimization of the network selection decision for users in order to support many services with best QoS and let the users stay connected with the current access network as long as possible. However, no single wireless network technology is considered to be more favorable than other technologies in terms of QoS. In other words, each network access in RAT's seems to be specifically characterized by the bandwidth offered, the coverage ensured by the network as well as the cost to deliver the service. Moreover, there is some kind of complementarity between these various networks, for example, 801.11a offers a higher bandwidth with a cover limited, while UMTS ensures a large cover with lower bandwidth. The network selection algorithm depends on multiple criteria which are:

- From terminal side: battery, velocity, etc.
- From service side: QoS level, security level, etc.
- From network side: provider's profile, current QoS parameters, etc.
- From user side: users preferences, perceived QoS, etc.

In the other hand the network selection problem can be tackled with several schemes and decision algorithms such as genetic algorithms [3], fuzzy logic [4], utility functions [5] and multi attribute decision making (MADM) methods [6,7,8,9,10,11,12]. In [3] the genetic algorithm is applied to optimize the access network function with the goal of selecting the optimal access network. In [4] the authors have proposed an intelligent approach for vertical handover based on fuzzy logic. In [5] the authors proposed a network selection scheme based on utility function which takes more key factors for multimedia communication in the future urban road wireless networks. These factors include data rate, bit error rate, latency, power consumption, monetary cost, load balance, individual's preference and handoff stability.

Due to great number of criteria and algorithms which can be used in network selection, the most challenging problems focus in selecting the appropriate criteria and definition of a strategy which can exploit these criteria. According to nature of network selection problem, MADM algorithms represent a promising solution to select the most suitable network in terms of quality of service (QoS) for mobile users. However the major limitations of MADM methods are the ranking abnormality and the ping-pong effect. The ranking abnormality means that the ranking of candidate networks change when low ranking alternatives are removed from the candidate list, which can make the selection problem inefficient. The ping pong effect occurs when the terminal mobile performs excessive handoffs for a given time which causing the higher number of handoffs. This phenomenon can led to increasing in power consumption and the decreasing in throughput.

To address the limitations posed by MADM methods, we propose an intelligent network selection strategy based on analytical network process (ANP) and the technique for order preference by similarity to an ideal solution (TOPSIS) method, the ANP method is applied to find the weights of each criterion and TOPSIS method is used to rank the alternatives. The intelligence of our strategy focuses in two aspects: firstly we utilize the differentiate weights of available networks by considering each criterion in order to reduce the ranking abnormality and secondly we introduce the history criterion to reduce the number of handoff and to ensure that the terminal mobile stay connected to the current access network as long as possible.

This paper is organized as follows. Section 2 presents review of related work concerning network selection decision based on MADM methods. Section 3 describes multi attribute decision making methods (MADM). Section 4 presents our access network selection algorithm based on ANP and TOPSIS two MADM methods. Section 5 includes the simulations and results. Section 6 concludes this paper.

2. RELATED WORK

The MADM methods represent promising solution for solving the network selection problem. The MADM includes many methods such as analytic hierarchy process (AHP), analytic network process (ANP), simple additive weighting (SAW), multiplicative exponential weighting (MEW), grey relational analysis (GRA), technique for order preference by similarity to ideal solution (TOPSIS) and the distance to the ideal alternative (DIA). In [6] comparison of network selection algorithms, between two methods which are the hybrid ANP algorithm and Blume algorithm is proposed. The hybrid ANP approach combines two MADM methods such as ANP method and rank reversal TOPSIS (RTOPSIS). The ANP method is used to get weights of the criteria and RTOPSIS method is applied to determine the ranking of access network. In [7] and [8], the network selection algorithm is based on AHP and GRA, the AHP method is used to determine weights for each criterion and GRA method is applied to rank the alternatives. In [9], [10] and [11] the network selection algorithm combines the AHP method and the TOPSIS method, the AHP method is used to get weights of the criteria and TOPSIS method is applied to determine the ranking of access network.

Among MADM methods mentioned above, TOPSIS method has been extensively used to solve the network selection problem. However, TOPSIS still suffers from ranking abnormality, some proposals were presented to avoid this issue, in [9] the author has proposed an iterative approach for application of TOPSIS for network selection problem. The disadvantage of this method lies in the computation time, for example, if we have n available access networks we must repeat iterative TOPSIS n-1 until the best interface network is reached. Reference [12] presents DIA algorithm which selects the alternative that is the shortest euclidean distance to positive ideal alternative. One of the main disadvantages of DIA method is doesn't take into account the normalization type, in other words, when the low ranking alternative is removed from the candidate list, the normalized attribute values of all alternatives will be changed and the ranking order of the alternative will be changed as well. Another disadvantage of this method is that, the euclidean distance used by DIA doesn't take into consideration the correlation between different criteria, all the components of the vectors will be treated in the same way.

The major factor causing the ranking abnormality is the weighting algorithm [13] used to weigh different criteria, in addition the all decision algorithms based on MADM methods use the same weight vector of the all available networks, in the other words each algorithm for network selection decision don't take into account the user preference relative to each access network according to each criterion. Due to the criteria are the same relative importance in each access network in the classical network selection algorithms, in our new strategy the ANP method is applied to find the differentiate weights of available networks by considering each criterion.

On the other hand the all selection decision algorithms based on MADM methods mentioned above still suffer from the ping-pong effect, to cope with this issue we introduce the history criterion to reduce the number of handoff and to ensure that the terminal mobile stay connected to the current access network as long as possible.

3. MULTI- ATTRIBUTE DECISION MAKING

3.1. ANP

The analytic network process (ANP) is a MADM method, proposed by Saaty [14], which extends the AHP approach to problems with dependence and feed beck within clusters (inner dependence) and between clusters (outer dependence). The ANP approach is based on six steps:

1) Model construction: A problem is decomposed into a network in which nodes corresponds to components. The elements in a component can interact with some or all of the elements of another component. Also, relationships among elements in the same component can exist. These relationships are represented by arcs with directions.

2) Construct of the pairwise comparisons: To establish a decision, ANP builds the pairwise matrix comparison such as:

$$A = \begin{bmatrix} x_{11} & x_{12} & \cdots & \cdots & x_{1n} \\ x_{21} & x_{22} & \cdots & \cdots & x_{2n} \\ \vdots & \vdots & \ddots & \vdots & \vdots \\ \vdots & \vdots & \vdots & \ddots & \vdots \\ x_{n1} & x_{n2} & \cdots & \cdots & x_{nn} \end{bmatrix}, where \begin{cases} x_{ii} = 1 \\ x_{ji} = \dfrac{1}{x_{ij}} \end{cases} \quad (1)$$

Elements x_{ij} are obtained from the table 1, it contains the preference scales.

Table 1: Saaty's scale for pair-wise comparison

Saaty's scale	The relative importance of the two sub-elements
1	Equally important
3	Moderately important with one over another
5	Strongly important
7	Very Strongly important
9	Extermely important
2,4,6,8	Intermediate values

3) Construct the normalized decision matrix: A_{norm} is the normalized matrix of $A(1)$, where $A(x_{ij})$ is given by, $A_{norm}(a_{ij})$ such:

$$A_{norm} = \begin{bmatrix} r_{11} & r_{12} & \cdots & \cdots & r_{1n} \\ r_{21} & r_{22} & \cdots & \cdots & r_{2n} \\ \vdots & \vdots & \ddots & \vdots & \vdots \\ \vdots & \vdots & \vdots & \ddots & \vdots \\ r_{n1} & r_{n2} & \cdots & \cdots & r_{nn} \end{bmatrix}, where \ r_{ij} = \frac{x_{ij}}{\sum_{i=1}^{n} x_{ij}} \quad (2)$$

4) Calculating the weights of criterion: The weights of the decision factor i can be calculated by:

$$W_i = \frac{\sum_{j=1}^{n} a_{ij}}{n} \ and \ \sum_{j=1}^{n} W_i = 1 \quad (3)$$

With n is the number of the compared elements.

5) Calculating the coherence ratio (CR): To test consistency of a pairwise comparison, a consistency ratio (CR) can be introduced with consistency index (CI) and random index (RI).

- Let define consistency index CI

$$CI = \frac{\lambda_{max} - n}{n - 1} \quad (4)$$

- Also, we need to calculate the λ_{max} by the following formula:

$$\lambda_{max} = \frac{\sum_{j=1}^{n} b_i}{n}, \qquad where \quad b_i = \frac{\sum_{j=1}^{n} W_i * a_{ij}}{W_i} \qquad (5)$$

- We calculate the coherence ratio CR by the following formula

$$CR = \frac{CI}{RI} \qquad (6)$$

The various values of RI are shown in table 2.

Table 2: value of random consistency index RI

Criteria	3	4	5	6	7	8	9	10
RI	0.58	0.90	1.12	1.24	1.32	1.41	1.45	1.49

If the CR is less than 0.1, the pairwise comparison is considered acceptable.

6) Construct the super-matrix formation: The local priority vectors are entered into the appropriate columns of a super-matrix, which is a partitioned matrix where each segment represents a relationship between two components.

3.2. TOPSIS

Technique for order preferences by similarity to an ideal solution (TOPSIS), known as a classical multiple attribute decision-making (MADM) method, has been developed in 1981 [15]. In TOPSIS method, the optimal alternative selected should have the shortest distance from the positive ideal solution and the farthest distance from the negative ideal solution. The procedure can be categorized in six steps:

1) Construct of the decision matrix: the decision matrix is expressed as

$$D = \begin{bmatrix} d_{11} & d_{12} & \cdots & \cdots & d_{1m} \\ d_{21} & d_{22} & \cdots & \cdots & d_{2m} \\ \vdots & \vdots & \ddots & \vdots & \vdots \\ \vdots & \vdots & \vdots & \ddots & \vdots \\ d_{n1} & d_{n2} & \cdots & \cdots & d_{nm} \end{bmatrix} \qquad (7)$$

Where d_{ij} is the rating of the alternative Ai with respect to the criterion C_j

2) Construct the normalized decision matrix: each element r_{ij} is obtained by the Euclidean normalization;

$$r_{ij} = \frac{d_i}{\sqrt{\sum_{i=1}^{m} d_{ij}^2}}, \text{ i=1,...,m and j=1,...,n.} \qquad (8)$$

3) Construct the weighted normalized decision matrix: The weighted normalized decision matrix v_{ij} is computed as:

$$v_{ij} = W_i * r_{ij} \quad where \quad \sum_{j=1}^{n} W_i = 1 \qquad (9)$$

4) Determination of the ideal solution A^* and the anti-ideal solution A^-:

$$A^* = [V_1^*, ..., V_m^*] \ and \ A^- = [V_1^-, ..., V_m^-] \qquad (10)$$

- For desirable criteria:

$$V_i^* = \max\{v_{ij}, i = 1, \dots, n\} \qquad (11)$$

$$V_i^- = \min\{v_{ij}, j = 1, \dots, n\} \qquad (12)$$

- For undesirable criteria:

$$V_i^* = \min\{v_{ij}, j = 1, \dots, n\} \qquad (13)$$

$$V_i^- = \max\{v_{ij}, j = 1, \dots, n\} \qquad (14)$$

5) Calculation of the similarity distance:

$$S_i^* = \sqrt{\sum_{j=1}^{m}(V_i^* - v_{ij})^2}, j = 1, \dots, n \qquad (15)$$

And

$$S_i^- = \sqrt{\sum_{j=1}^{m}(V_i^- - v_{ij})^2}, j = 1, \dots, n \qquad (16)$$

6) Ranking:

$$C_j^* = \frac{S_j^-}{S_i^* + S_i^-}, j = 1, \dots, n \qquad (17)$$

A set of alternatives can be ranked according to the decreasing order of C_j^*.

4. ACCESS NETWORK SELECTION STRATEGY

In order to deal with the ranking abnormality and to reduce the number of handoffs, we propose new intelligent network selection strategy based on two MADM methods such as ANP method and TOPSIS method. The ANP method is applied to find the weights of available networks by considering each criterion and TOPSIS method is applied to determine the ranking of each access network. Moreover our strategy introduces a new criterion namely history. This attribute allows to memorise the overall score given to the available network by using the TOPSIS method (history value is C_j^*).

The algorithm assumes wireless overlay networks which entail three heterogeneous networks such as UMTS, WLAN and WIMAX. Instead of using six attributes associated in this heterogeneous environment which are: Cost per Byte (CB), Available Bandwidth (AB), Security (S), Packet Delay (D), Packet Jitter (J) and Packet Loss (L), we add a new history criterion (H). Due to relationships between the QoS parameters such as AB, D, J and L, and based on survey and comparison study on weighting algorithms for access network selection presented in [13], the ANP method is the most appropriate algorithm which can be used to assign weights for each criterion.

Figure 1. exhibits the three levels based on ANP hierarchy for our new network selection strategy which takes into consideration the history attribute. The level 1 includes four criteria QoS, security, cost and history, the level 2 includes four QoS parameters such as AB, D, J and L and the level 3 includes three available networks UTMS, WIFI and WIMAX.

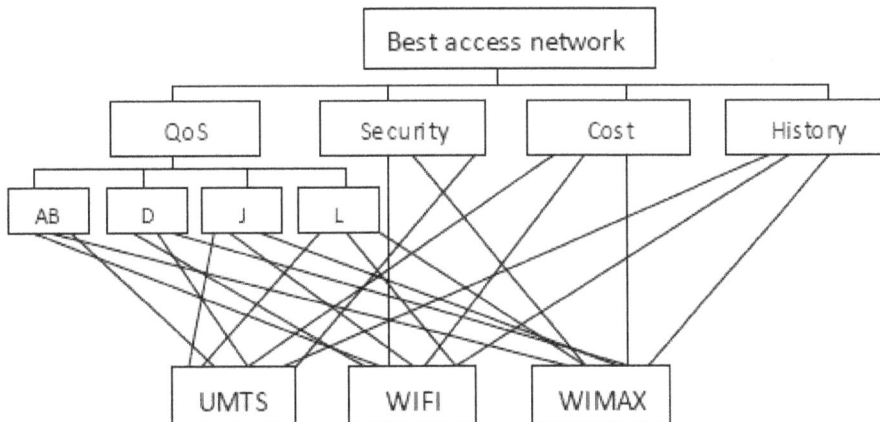

Figure 1. ANP hierarchy for our network selection problem

Based on the specific characteristics of the traffic type [16], our new strategy can be categorized in five steps:

1) Assign weights to level-1-criteria: the ANP method is used to get a weight of the decision criteria of level 1.
2) Assign weights to level-2-criteria: the ANP method is used to get a weight of the decision criteria of level 2 and to eliminate the interdependence impact of QoS sub-criteria.
3) Assign weights to level-3-alternatives: the ANP method is used to find the weights of the available networks by considering each criterion.
4) Obtain the vector weights of each available network: each access network will have dissimilar unique weights vector which will differ from those of other available networks, the weight vector of each available network is calculated by multiplication of the weight vector obtained in level 1 with the weight vector obtained in level 2 and with the weight vector obtained in level 3.
5) Select the best access network: the method TOPSIS is applied to rank the available networks and select the access network that has the highest value of C_j^* (see the steps of TOPSIS method).

5. SIMULATION AND RESULTS

In order to illustrate the effectiveness of our new strategy based on ANP and TOPSIS which taking into consideration the user preference relative to each access network according to each criterion and including a new history attribute, we present performance comparison between four algorithms namely:

- TOPSIS-1: the network selection algorithm combines ANP method and TOPSIS method without considering differentiated weight of criterion and without considering the history attribute.
- TOPSIS-2: the network selection approach is based on ANP and TOPSIS and taking on consideration only the history attribute.
- TOPSIS-3: the network selection algorithm is based on ANP and TOPSIS and taking on consideration only the differentiated weight of criterion according to specific access network.

- TOPSIS-4: it's our new network selection strategy based on ANP and TOPSIS which considering differentiated weight of criterion according to specific access network and including the history attribute.

We simulate four traffic classes [16] namely background, conversational, interactive and streaming. In each simulation the four algorithms were run in 12 vertical handoff decision points and the performance evaluation is focused on two aspects, which are ranking abnormality and number of handoffs. For TOPSIS-1 and TOPSIS-3 the history criterion for each access network has no effect on our simulation.

Table 3: Attribute values for the candidate networks

Criteria Network	CB (%)	S (%)	AB (mbps)	D (ms)	J (ms)	L (per10^6)	H (%)
UMTS	60	70	0.1-2	25-50	5-10	20-80	100
WLAN	10	50	1-11	100-150	10-20	20-80	100
WIMAX	40	60	1-60	60-100	3-10	20-80	100

During the simulation, for each candidate networks, the measures of six attributes CB, AB, S, D, J and L are randomly varied according to the ranges shown in table 3. Furthermore the value of history criterion is initialized by 1, after the value of H_{i+1} is equal to C_j^* in iteration i+1 where C_j^* is the score of TOPSIS method obtained in iteration i.

5.1. Simulation 1

In this simulation, the traffic analyzed is background traffic, the weight vector of TOPSIS-1 and TOPSIS-2 are displayed in figure 2 and the weight vector of each network such as WIFI, WIMAX, and UMTS which calculated by TOPSIS-3 and TOPSIS-4 are displayed in figure 3 and figure 4 respectively.

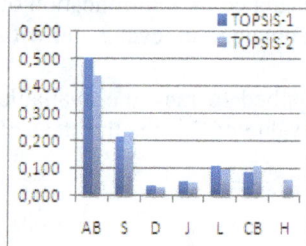

Figure 2. Weights of TOPSIS-1 and TOPSIS-2

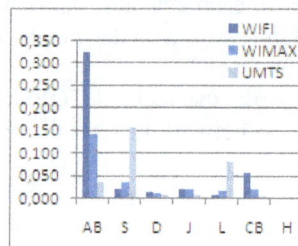

Figure 3. Weights of TOPSIS-3

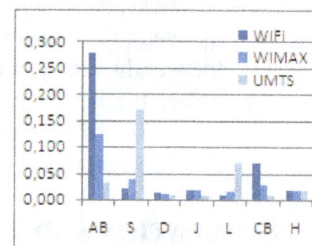

Figure 4. Weights of TOPSIS-4

5.1.1. Ranking abnormality

Figure 5. shows that TOPSIS-1 method reduces the risk to have this problem with a value of 33%, and TOSIS-2 method and TOPSIS-3 provide the same value for reducing the risk with a value of 25%. While TOPSIS-4 method reduces the risk with a value of 8%.

So for background traffic, TOPSIS-4 method based on differentiated weight and history attribute can reduce the ranking abnormality problem better than the all algorithms such as TOSIS-1, TOPSIS-2 and TOSIS-3, in addition the TOPSIS-2 and TOPSIS-3 which taking into consideration the differentiate weight and history attribute respectively reduce the ranking abnormality problem better than the classical network selection based on TOPSIS-1.

5.1.2. Number of handoffs

Figure 6. shows that TOPSIS-1 method reduces the number of handoffs with a value of 42%, and TOPSIS-2 method and TOPSIS-3 method provide the same value of the number of handoffs, the value is 25%. While TOPSIS-4 method reduces the number of handoffs with a value of 8%.

So for background traffic, TOPSIS-4 method based on differentiated weight and history attribute can reduce the number of handoffs better than the all algorithms such as TOPSIS-1, TOPSIS-2 and TOPSIS-3, in addition the TOPSIS-2 and TOPSIS-3 which taking into consideration the differentiate weight and history attribute respectively reduce the number of handoffs better than the classical network selection based on TOPSIS-1.

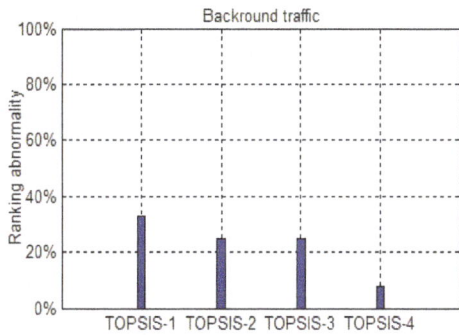

Figure 5. Average of ranking abnormality

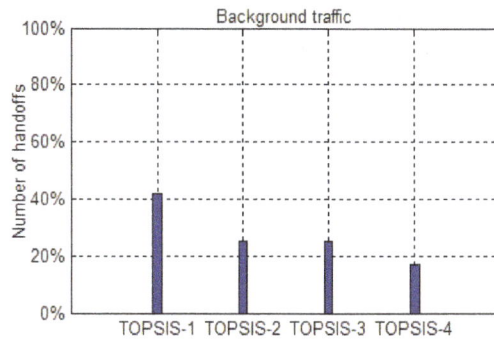

Figure 6. Average of number of handoffs

5.2. Simulation 2

In this simulation, the traffic analyzed is conversational traffic, the weight vector of TOPSIS-1 and TOPSIS-2 are displayed in figure 7 and the weight vector of each network such as WIFI, WIMAX, and UMTS which calculated by TOPSIS-3 and TOPSIS-4 are displayed in figure 8 and figure 9 respectively.

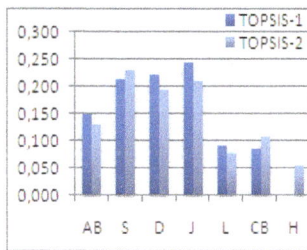

Figure 7. Weights of TOPSIS-1 and TOPSIS-2

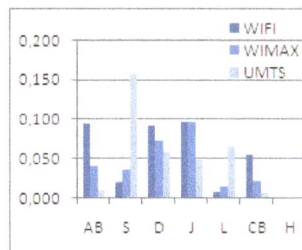

Figure 8. Weights of TOPSIS-3

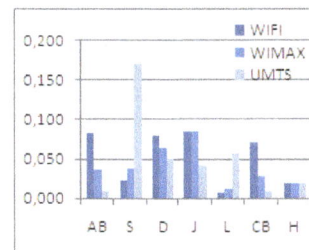

Figure 9. Weights of TOPSIS-4

5.2.1. Ranking abnormality

Figure 10. shows that TOPSIS-1 method reduces the risk to have this problem with a value of 25%, and TOSIS-2 method and TOPSIS-3 provide the same value for reducing the risk with a value of 17%. While TOPSIS-4 method reduces the risk with a value of 8%.

So for conversational traffic, TOPSIS-4 method based on differentiated weight and history attribute can reduce the ranking abnormality problem better than the all algorithms such as TOPSIS-1, TOPSIS-2 and TOPSIS-3, in addition the TOPSIS-2 and TOPSIS-3 which taking into consideration the differentiate weight and history attribute respectively reduce the ranking abnormality problem better than the classical network selection based on TOPSIS-1.

5.2.2. Number of handoffs

Figure 11. shows that TOPSIS-1 method reduces the number of handoffs with a value of 50%, and TOPSIS-2 method and TOPSIS-3 method provide the same value of the number of handoffs, the value is 42%. While TOPSIS-4 method reduces the number of handoffs with a value of 8%.

So for conversational traffic, TOPSIS-4 method based on differentiated weight and history attribute can reduce the number of handoffs better than the all algorithms such as TOPSIS-1, TOPSIS-2 and TOPSIS-3, in addition the TOPSIS-2 and TOPSIS-3 which taking into consideration the differentiate weight and history attribute respectively reduce the number of handoffs better than the classical network selection based on TOPSIS-1.

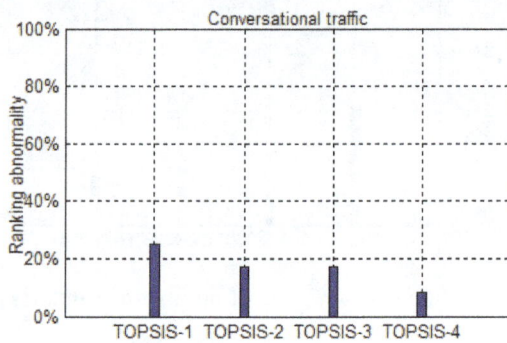

Figure 10. Average of ranking abnormality   Figure 11. Average of number of handoffs

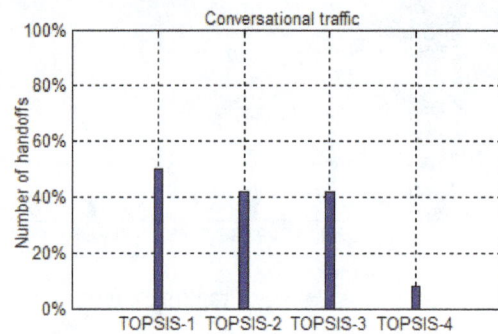

5.3. Simulation 3

In this simulation, the traffic analyzed is interactive traffic, the weight vector of TOPSIS-1 and TOPSIS-2 are displayed in figure 12 and the weight vector of each network such as WIFI, WIMAX, and UMTS which calculated by TOPSIS-3 and TOPSIS-4 are displayed in figure 13 and figure 14 respectively.

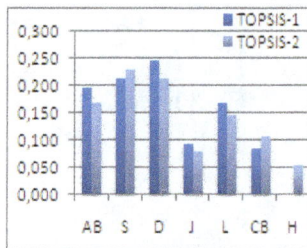

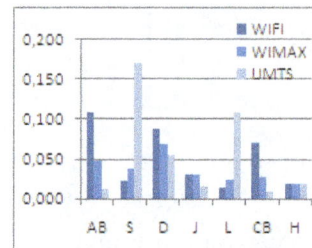

Figure 12. Weights of TOPSIS   Figure 13. Weights of   Figure 14. Weights of
1 and TOPSIS-2   TOPSIS-3   TOPSIS-4

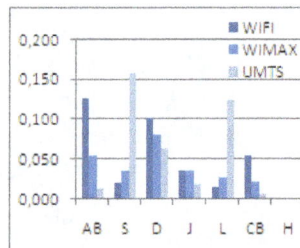

5.3.1. Ranking abnormality

Figure 15. shows that TOPSIS-1 method reduces the risk to have this problem with a value of 25%, and TOSIS-2 method and TOPSIS-3 provide the same value for reducing the risk with a value of 17%. While TOPSIS-4 method reduces the risk with a value of 8%.

So for interactive traffic, TOPSIS-4 method based on differentiated weight and history attribute can reduce the ranking abnormality problem better than the all algorithms such as TOPSIS-1, TOPSIS-2 and TOPSIS-3, in addition the TOPSIS-2 and TOPSIS-3 which taking into consideration the differentiate weight and history attribute respectively reduce the ranking abnormality problem better than the classical network selection based on TOPSIS-1.

5.3.2. Number of handoffs

Figure 16. shows that TOPSIS-1 method reduces the number of handoffs with a value of 33%, and TOPSIS-2 method and TOPSIS-3 method provide the same value of the number of handoffs, the value is 25%. While TOPSIS-4 method reduces the number of handoffs with a value of 8%.

So for interactive traffic, TOPSIS-4 method based on differentiated weight and history attribute can reduce the number of handoffs better than the all algorithms such as TOPSIS-1, TOPSIS-2 and TOPSIS-3, in addition the TOPSIS-2 and TOPSIS-3 which taking into consideration the differentiate weight and history attribute respectively reduce the number of handoffs better than the classical network selection based on TOPSIS-1.

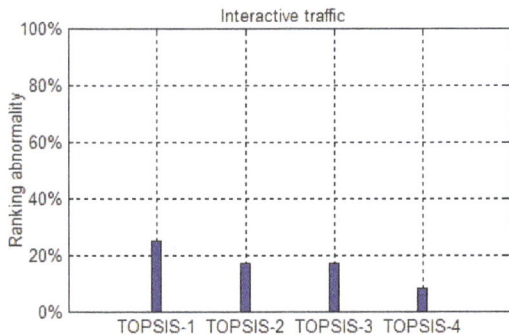

Figure 15. Average of ranking abnormality   Figure 16. Average of number of handoffs

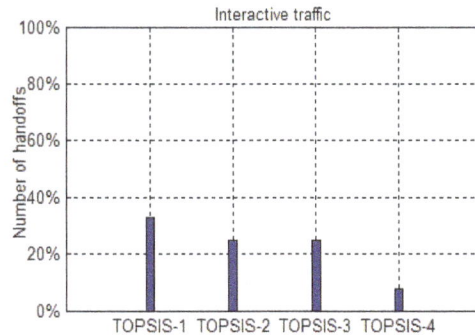

5.4. Simulation 4

In this simulation, the traffic analyzed is streaming traffic, the weight vector of TOPSIS-1 and TOPSIS-2 are displayed in figure 17 and the weight vector of each network such as WIFI, WIMAX, and UMTS which calculated by TOPSIS-3 and TOPSIS-4 are displayed in figure 18 and figure 19 respectively.

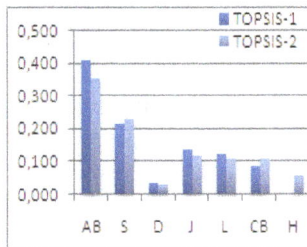

Figure 17. Weights of TOPSIS-1 and TOPSIS-2

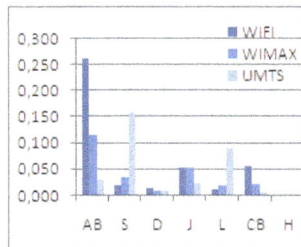

Figure 18. Weights of TOPSIS-3

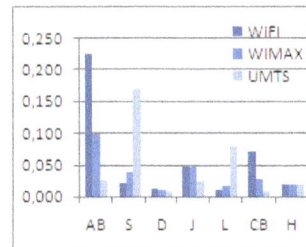

Figure 19. Weights of TOPSIS-4

5.4.1. Ranking abnormality

Figure 20. shows that TOPSIS-1 method reduces the risk to have this problem with a value of 42%, TOPSIS-2 method reduces the risk with a value of 33% and TOPSIS-3 method reduces the risk with a value of 25%. While TOPSIS-4 method reduces the risk with a value of 17%.

So for streaming traffic, TOPSIS-4 method based on differentiated weight and history attribute can reduce the ranking abnormality problem better than the all algorithms such as TOPSIS-1, TOPSIS-2 and TOPSIS-3, in addition the TOPSIS-2 and TOPSIS-3 which taking into consideration the differentiate weight and history attribute respectively reduce the ranking abnormality problem better than the classical network selection based on TOPSIS-1.

5.4.2. Number of handoffs

Figure 21. shows that TOPSIS-1 method reduces the number of handoffs with a value of 58%, TOPSIS-2 method reduces the number of handoffs with a value of 50% and TOPSIS-3 method reduces the number of handoffs with a value of 33%. While TOPSIS-4 method reduces the number of handoffs with a value of 25%.

So for streaming traffic, TOPSIS-4 method based on differentiated weight and history attribute can reduce the number of handoffs better than the all algorithms such as TOPSIS-1, TOPSIS-2 and TOSIS-3, in addition the TOPSIS-2 and TOPSIS-3 which taking into consideration the differentiate weight and history attribute respectively reduce the number of handoffs better than the classical network selection based on TOPSIS-1.

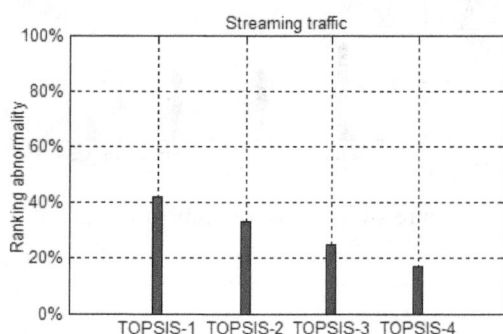

Figure 20. Average of ranking abnormality

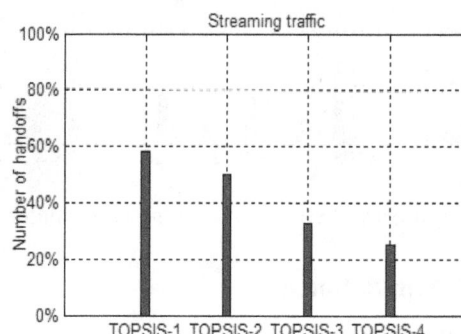

Figure 21. Average of number of handoffs

6. CONCLUSIONS

In this work, we have proposed an intelligent network selection strategy namely TOPSIS-4. This strategy combines two MADM methods such as ANP method and TOPSIS method, the ANP method is applied to find the differentiate weights of available networks by considering each criterion and the TOPSIS method is used to rank the available networks. In addition the proposed strategy takes into consideration a new attribute namely history. This one helps to deal with the ping pong effect by reducing the number of handoffs.

The simulation results show that, our method based on TOPSIS-4 can reduce the ranking abnormality problem better than all algorithms such as TOPSIS-1, TOPSIS-2 and TOPSIS-3 according to all four traffic classes namely background, conversational, interactive and streaming. In the other hand for all traffic classes TOPSIS-4 method provides best performance concerning the number of handoffs than TOPSIS-1, TOPSIS-2 and TOPSIS-3.

Finally we deduce that the introducing of the differentiated weight (TOPSIS-2) or the history criterion (TOPSIS-3) in the network selection decision allows to get the best performance concerning the two aspects namely ranking abnormality and number of handoffs than the classical network selection decision (TOPSIS-1).

REFERENCES

[1]   E. Gustafsson and A. Jonsson, "Always best connected", IEEE Wireless Communications Magazine, vol.10, no.1,pp.49-55, Feb. 2003.

[2]   H. Wang, R. Katz, J. Giese, Policy-enabled handoffs across heterogeneous wireless networks, Second IEEE Worshop on Mobile Computing systems and Applications, WMCSA. pp. 51-60, February 1999.

[3]   Nkansah-Gyekye, Y. Agbinya, J.I. "A Vertical Handoff Decision Algorithm for Next Generation Wireless Networks", Third International Conference on Broadband Communications, Information Technology and Biomedical Applications, pp.358-364, Nov. 2008.

[4]   H. Attaullah et al, Intelligent vertical handover decision model to improve QoS, In Proceedings of ICDIM, pp.119-124, 2008.

[5]   L. Chen et al, An utility-based network selection scheme for future urbanroad wireless networks, In the 5th IEEE Conference on Industrial Electronics and Applications (ICIEA), pp.181-185, June 2010.

[6]   R. Kumar, B. Singh. "Comparison of vertical Handover Mechanisms Using Generic QoS Trigger for Next Generation Network", In the International Journal of Next-Generation Networks (IJNGN) Vol.2, N.3, pp.80-97, Septembre 2010.

[7]   J. Fu, J. Wu, J. Zhang, L. Ping, and Z. Li, " Novel AHP and GRA Based Handover Decision Mechanism in Heterogeneous Wireless Networks", in Proc. CICLing (2), pp.213-220, 2010.

[8]   Wang Yafang; Cui Huimin; Zhang Jinyan. "Network access selection algorithm based on the Analytic Hierarehy Process and Gray Relation Analysis", in Proc. New Trends in Information Science and Service Science NISS'2010, pp.503-506, May 2010.

[9]   F. Bari and V. Leung, "Multi-attribute network selection by iterative TOPSIS for heterogeneous wireless access", 4th IEEE Consumer Communications and Networking Conference, pp.808-812, January 2007.

[10]   A. Sgora, D. Vergados, P. Chatzimisios. "An access network selection algorithm for heterogeneous wireless environments", iscc, pp.890-892, The IEEE symposium on Computers and Communications, 2010.

[11]   M. Lahby, C. Leghris. and A. Adib. "A Hybrid Approach for Network Selection in Heterogeneous Multi-Access Environments", In the Proceedings of the 4th IFIP International Conference on New Technologies, Mobility and Security (NTMS),pp. 1-5, 2011.

[12]   P.N. Tran and N. Boukhatem. "The distance to the ideal alternative(DiA) algorithm for interface selection in heterogeneous wireless networks", Proceedings of The 6th ACM international symposium on Mobility management and wireless access (MobiWac08), Pages 61-68, Oct. 2008

[13]   M. Lahby, C. Leghris. and A. Adib. "A Survey and Comparison Study on Weighting Algorithms for Access Network Selection", In the Proceedings of the 9th Annual Conference on Wireless On-Demand Network Systems and Services (WONS), pp.35-38, January 2012 (to appear).

[14]   J. Lee, and S. Kim, "Using Analytic Network Process and Goal Programming for Interdependent Information System Project Selection", Computers and Operation Research, Volume 27, Number 4, Page 367-382, April 2000.

[15]   E. Triantaphyllou "Multi-Criteria Decision Making Methods: A Comparative Study", Kluwer academic publishers, Applied optimization series, Vol. 44, 2002.

[16]   "3GPP, QoS Concepts and Architecture" 2005, tS 22.107 (v 6.3.0).

2

USER CENTRIC NETWORK SELECTION IN WIRELESS HETNETS

L.Nithyanandan[1], V.Bharathi[2] and P.Prabhavathi[3]

[1,2,3]Department of Electronics and Communication Engineering,
Pondicherry Engineering College, Puducherry, India
nithi@pec.edu
bharathime@rediffmail.com
prabhatech@pec.edu

ABSTRACT

The future generation wireless networks are expected to beheterogeneous networksconsisting of UMTS, WLAN, WiMAX, LTE etc. A heterogeneous network provides users with different data rate and Quality of Service (QoS). Users of future mobile networks will be able to choose from different radio access technologies. These networks vary widely in service capabilities such as coverage area, bandwidth and error characteristics. Network selection is a challenging task in heterogeneous networks and will influence the performance metrics of importance for both service provider and subscriber.This paper analysesuser centric network selection based on QoE (Quality of Experience) which include both technical and economical aspects of the user. WLAN-WiMAX-UMTS networks are integrated and the network selection for the integrated network is performed using game theory based network selection algorithm.

KEYWORDS

UMTS, WLAN, WiMAX, QoS, QoE and Heterogeneous network

1. INTRODUCTION

There is a rapid development in the wireless communication nowadays because there is an increase in number of mobile and internet users. One such new technology is the heterogeneous network which refers to the integration of different Radio Access Technology (RAT) such as WLAN, WiMAX, UMTS etc. These wireless networks differ in its bandwidth, coverage area, data rate, mobility, technology etc. WLAN network is basically 802.11 standardwhich provides less coverage and high data rates (54Mbps). WiMAX is 802.16 standard with data rates up to 75Mbps and coverage nearly 30Km. UMTS is 3GPP standard which offers less data rate (2Mbps) and more coverage.Heterogeneous network provides benefits such as seamless connectivity, ubiquitous availability of multimedia services and effective utilization of available bandwidth to meet user requirements. In a wireless environment user's mobility, characteristics and availability of a network will change in time. So, the mobility management process considers the dynamic reselection of network as its major task.Network selection is necessary when users want to migrate between heterogeneous networks. Hence multi mode mobile terminal is used to select a network from the converged network.

Network selection is a quite difficult process in wireless environment since network condition varies randomly based on variation in user demands, random activity of users and vagueness of system parameters. In case of network centric approach, base station consists of a centralized controller which admits the user to a particular network and allocates bandwidth to the user to maximize network utility. The user's needs and optimum results are not achieved through this approach. Hence, to have a better performance, network selection can be done using user centric approach. In user centric approach, game theory based network selection algorithms are implemented at the user terminal. Future heterogeneous network will be consisting of different RANs at same place and network conditions vary according to channel condition,random requirement of users etc. In this realistic environment network selection using Game theory produces optimum result for users.

The rest of this paper is structured as follows. Section 2 gives the literature survey. Network selection process is explained in section 3. Section 4 describes the network selection decision making. Mapping of game theory to network selection is given in section 5. Section 6 explains the simulation results and finally section 7 concludes the paper.

2. LITERATURE SURVEY

J.Park, *et al.,* [1] has proposed the user subscription dynamics, revenue maximization and equilibrium characteristics in two different markets (i.e., monopoly and duopoly). A fuzzy-logic-based multiple-criteria decision making system to perform access network selection in a heterogeneous network environment is described in [2]. A simple policy enabled handoff across heterogeneous wireless networks is presented in [3], which allows users to express policies which selects best wireless network at any moment, and that makes trade-offs among network characteristics and dynamics such as cost, performance and power consumption.A network-selectionstrategy that only considers mobile users power consumption was introduced by M.Nam*et.al* [4]. Another network selection algorithm was proposed by Q.Y.Song*et.al* [5] in which Analytical Hierarchy Process (AHP) and the Grey Relational Analysis (GRA) were discussed. The AHP algorithm divides the complex network-selection problem into a number of decision factors, and the optimal solution can be found by integrating the relative dominance among these factors. The GRA was also proposed for selecting the best network for a mobile user. This mathematical network selection model is suitable for static environment. Later in 2012 Manzoor Ahmed Khan *et.al,*[6] introduce da dynamic and uncertain environment for network selection where the users and operators have only a numerical value of their own payoffs as information, and constructed various heterogeneous COmbined fully Distributed Payoff and Strategy Reinforcement Learning (CODIPAS-RL) in which the users try to learn their own optimal payoff and their optimal strategy simultaneously. In [6] Bush Mosteller based CODIPAS-RL for interworking two networks is discussed. In this paper Bush Mosteller and less iterative Boltzmann-Gibbs CODIPAS-RL are discussed and it is implemented in three network interworked architecture simulated using OPNET.

3. NETWORK SELECTION PROCESS

Network selection process is defined as the process of selecting a specific network from the converged network which meets the user's requirements. To introduce the concept of Always Best Connected (ABC) anytime and anywhere, network selection mechanisms are facilitated in heterogeneous networks. Different Radio Access Networks (RAN) differs in coverage, data rate, QoS, pricing scheme etc., hence it is difficult for a user to select a particular network [7]. From the user's perspective, the variety of portable devices (such as smart phones, net books, or laptops) with support for multiple radio network interfaces, enable the option of connecting to the Internet anywhere and anytime. Users are able to freely migrate from one Radio Access Technology (RAT) to another or from one service provider to another. A network selection decision is made at call setup and subsequently the decision is re-made in the case of a handover trigger. The major phases in network selection process are Monitoring, Network selection and Call setup. In monitoring phase the mobile terminal monitors the network conditions, lists the available RAN's and predicts the characteristics of each RAN. By using these monitored data the mobile terminal triggers a Hand Off (HO) decision.

In network selection phase network selection process is initiated either by automatic trigger for HO for existing call or by a request for a new connection [8]. The best network is decided based on decision criteria provided by the device, application and monitoring process. When the target network is selected the call is set up on the candidate network. But conventionally this decision was made by network operators mainly based on Received Signal Strength. When the target network is selected, the connection set up on the target candidate network is executed. But in case of an existing network connection, HO is executed and the original connection is removed and the call is re-routed to the new connection. If first choice network is unavailable, then the next listed candidate is chosen as target network. Connection setup and connection release are handled by Mobile Internet Protocol Version 6.

4. NETWORK SELECTION DECISION MAKING

Every decision making mechanism requires essential and relevant input information in order to choose the best network [9]. The decision criteria that may be used in the network selection process are network metrics, device related information, application requirements and user preferences. Network metrics includes information about the technical characteristics or performance of the access networks such as technology type, coverage, security, pricing scheme, monetary cost, available bandwidth, network load, latency, Received Signal Strength, blocking probability, network connection time, etc. Device related information refers to information about the end-users' terminal device characteristics like supported interfaces, mobility support, capacity, capability, screen-size and resolution, location-based information, remaining battery power, etc. Application requirement refers to information about the requirements (minimum and maximum thresholds) needed in order to provide a certain service to the end user: delay, jitter, packet loss required throughput, Bit Error Rate, etc. User preferences comprise information related to the end-users satisfaction such as budget (willingness to pay), service quality expectations, energy conservation needs, etc. The user preferences play an important role in the decision mechanism and they may be used to weight the other parameters involved.

5. MAPPING OF GAME THEORY TO NETWORK SELECTION

Game theory is a mathematical tool used in understanding and modelling competitive situations. In the wireless environment, game theory has been used in order to solve many distributed power control, resource management and dynamic pricing related problems[10].The main components of a game are: the set of players, the set of actions, and the set of payoffs. The players seek to maximize their payoffs by choosing strategies that deploy actions depending on the available information at a certain moment. Each player chooses strategies which can maximize their payoff. The combination of best strategies for each player is known as equilibrium. The mapping of game theory to network selection is given in Table 1.

Table 1. Mapping of game theory to network selection

Game Component	Network Selection Environment Correspondent
Players	The agents who are playing the game : Users or networks
Strategies	A plan of actions to be taken by the player during the game: available/ requested bandwidth, subscription plan, offered prices, available APs. etc.
Payoffs	The motivation of players represented by profit and estimated using utility functions based on various parameters: monetary cost, quality, network load, QoS, etc.
Resources	The resources for which the players involved in the game are competing: bandwidth, power, etc.

6. NETWORK SELECTION ALGORITHMS

There are different strategies in network selection such as select the economic RAN, random network selection, select the preferred operator's network. But these strategies do not provide optimum results in a dynamic wireless network. An intelligent approach for network selection provides optimum results. Hence, to get optimum results network selection algorithm based on game theory was introduced[11]. COmbined and fully Distributed Payoff and Strategy Reinforcement Learning (CODIPAS-RL) were introduced in [6]. In this paper UMTS-WIMAX-WLAN interworked environment is simulated. Bush-Mosteller based and Boltzmann-Gibbs based CODIPAS-RL is implemented for UMTS-WIMAX-WLAN interworked environment.

6.1. Bush Mosteller based CODIPAS-RL

Bush Mosteller based CODIPAS-RL is a stochastic model of reinforcement learning where users decide which action to take stochastically: each user's strategy is defined by the probability of undertaking each of the two actions available to them. After every user has selected an action according to their probability, every user receives the corresponding utility and revises their strategy [6, 11, 12]. The action of each user is determined based on (1)

$$x_{j,t+1}(a_j) = x_{j,t}(a_j) + \lambda j S_{j,t}(1 - x_{j,t}(a_j)) \tag{1}$$

Where λ_j is the user j's learning rate ($0 < \lambda < 1$) and $S_{j,t}$ denotes stimulus of user which is given in (2) and its values ranges from [-1, 1]. When stimulus magnitude or learning rate increases, then the change in probability also increase.

$$S_{j,t} = \frac{u_{j,t} - M_j}{sup_a |U_j(a) - M_j|} \tag{2}$$

where $u_{j,t}$ denotes the perceived utility at time t of player j and M_j is an aspiration level of player j. The payoff estimation for the experimented actions by the users is given in the eq. (3).

$$\hat{u}_{j,t+1}(a_j) = \hat{u}_{j,t}(a_j) + v(t) * (u_{j,t} - \hat{u}_{j,t}(a_j)) \tag{3}$$

Bush Mosteller based CODIPAS-RL considers present action of user j as well as the actions of other users. So it requires more memory and processing time is also high. Hence the number of iterations or time required for network selection process is more.

6.2. Boltzmann-Gibbs based CODIPAS-RL

Since Bush Mosteller based CODIPAS-RL takes more number of iterations for network selection, Boltzmann-Gibbs based CODIPAS-RL was used because it considers the previous action and present action of user j alone and requires less memory. So, less number of iterations only are required in this algorithm [13, 14]. The action of the user is based on Boltzmann distribution which is given in [6,11] as

$$B_{j,\in}(\hat{u}_{j,t})(a_j) = \frac{e^{\frac{1}{\in_j}\hat{u}_{j,t}(a_j)}}{\sum_{a'_j \in A_j} e^{\frac{1}{\in_j}\hat{u}_{j,t}(a'_j)}}, a_j \in A_j, j \in K \tag{4}$$

Where $\frac{1}{\epsilon_j}$ denotes the rationality level of the user j. $\hat{u}_{j,t}$ denotes the estimated payoff of user j at time t and a_j denotes the action of user j at time t. Based on action of user in (4), the payoff for each user is calculated using eqn. (5).

$$\hat{u}_{j,t+1}(a_j) = \hat{u}_{j,t}(a_j) + v(t) * (u_{j,t} - \hat{u}_{j,t}(a_j)) \tag{5}$$

The steps for selecting a network are given as follows:

Step 1: Assume users are initially connected to a network and calculate the payoff in current network

Step 2: Check for the alternate networks where users can connect

Step 3: Determine the payoff and action of the users based on Boltzmann-Gibbs CODIPAS-RL

Step 4: Determine Network Selection Probability (NSP) based on the action of users

7. SIMULATION RESULTS

For effective network selection process three networks such as WLAN, WiMAX and UMTS are integrated using OPNET modeller [15] and the network selection process is carried out for users using MATLAB.Fig.1 shows the architecture of WLAN-UMTS-WiMAX integration. The architecture is based on hybrid coupling in which the data packets between WLAN and WiMAX reaches the UMTS network directly through Gateway GPRS Support Node (GGSN) or through the IP core network based on data type(real time or non real time). The IEEE (WLAN, WiMAX) and 3GPP (UMTS) networks are integrated using IMS (IP Multimedia Subsystem) which is based on SIP (Session Initiation Protocol) and terminal mobility is provided using MIPV6. SIP is an application layer protocol which is used to establish, maintain and terminate the session. The basic components of SIP are P-CSCF (Proxy Call Session Control Function), I-CSCF (Interrogating Call Session Control Function) S-CSCF (Serving Call Session Control Function) and Home Subscriber Station (HSS). Within the coverage of UMTS two networks (WLAN and WiMAX) are present and within the coverage of WiMAX, WLAN is present. The architecture consists of three users with three different applications such as video, voice and web browsing.

Figs.2, 3 and 4 show the throughput, delay and load in WLAN network. The maximum throughput of the network is found to be 0.2 Mbps this is because only less number of users is considered. Throughput of the network increases as the number of users increase.

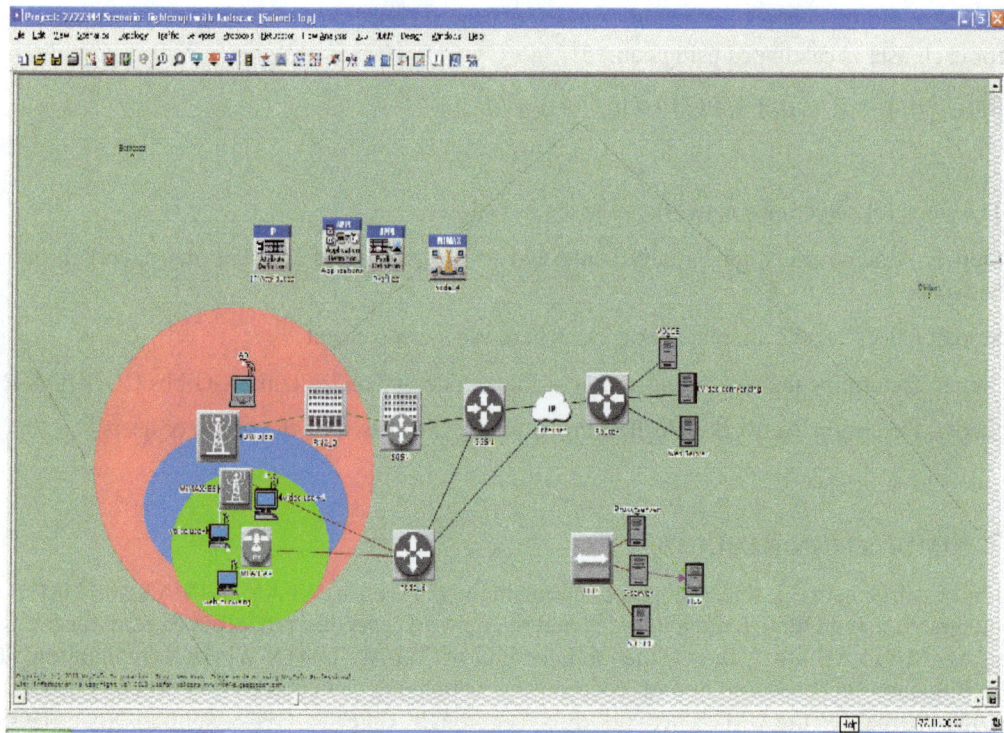

Fig.1 WLAN-WiMAX-UMTS architecture

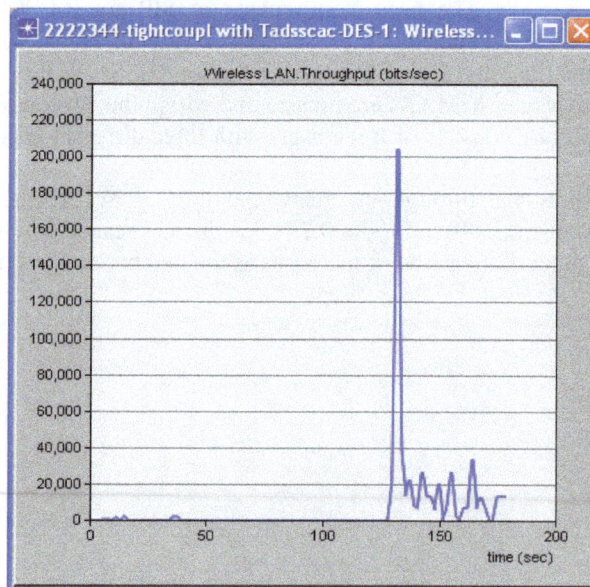

Fig.2 Throughput in WLAN

Figs.5, 6 and 7 show the throughput, delay and load in UMTS network. Since the number of users considered in the architecture is minimum, the throughput obtained is less compared to the actual network throughput. The network throughput decreases based on loading conditions also. The delay in UMTS is less than WLAN network because UMTS is a circuit switched network whereas WLAN is a packet switched network.

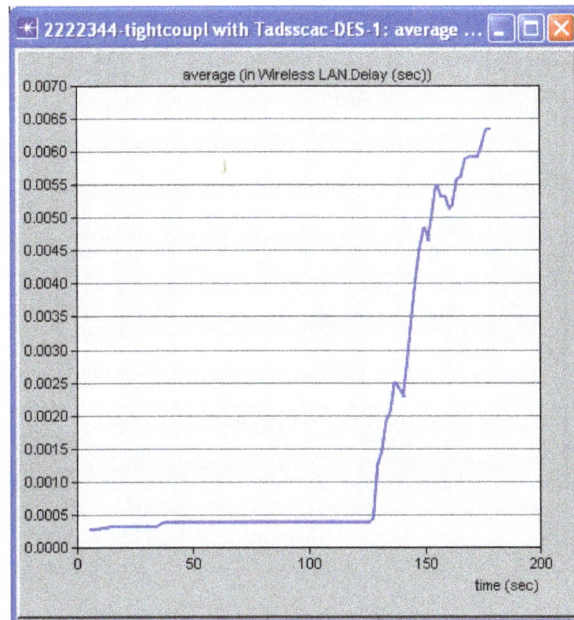

Fig.3 Delay in WLAN

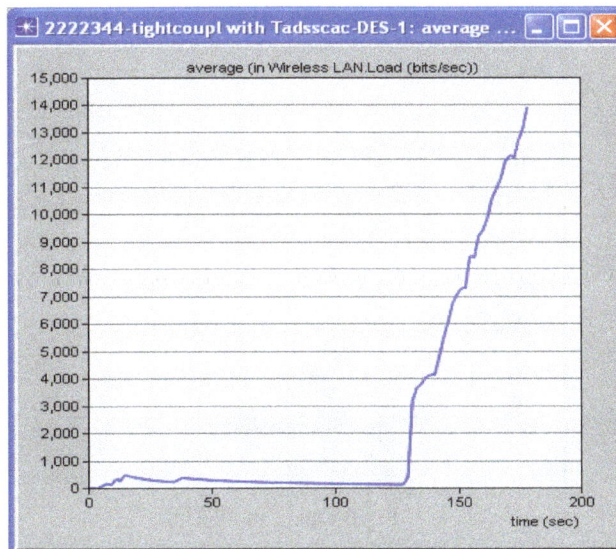

Fig. 4 WLAN Load

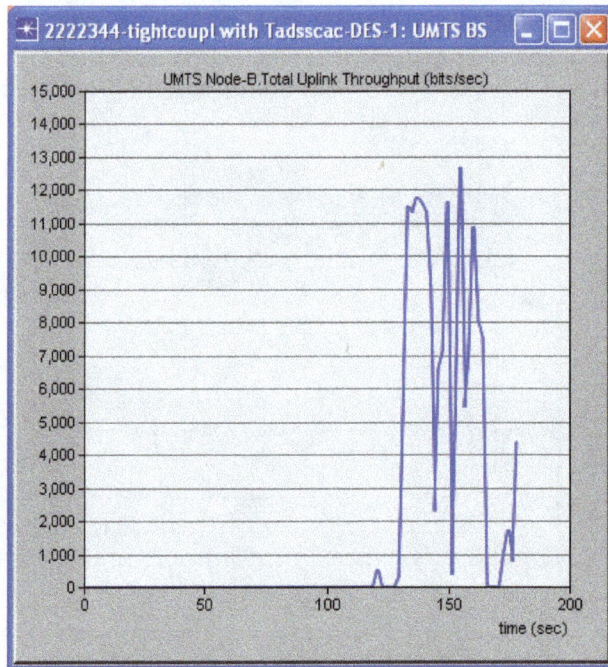

Fig.5 Throughput in UMTS

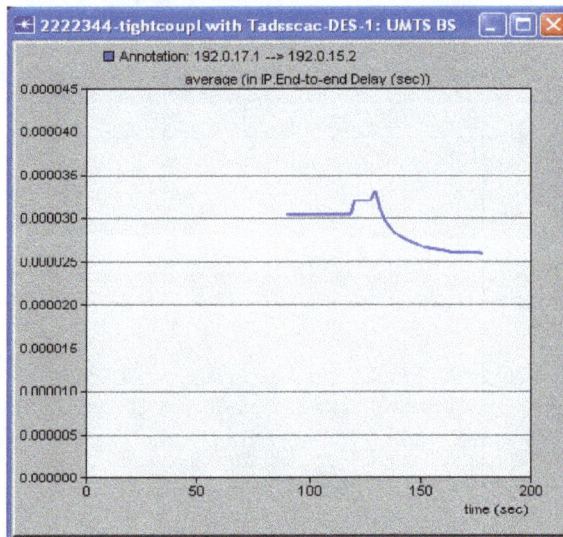

Fig.6 Delay in UMTS

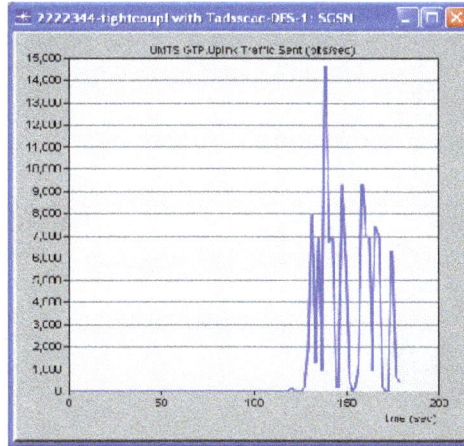

Fig.7 Load in UMTS

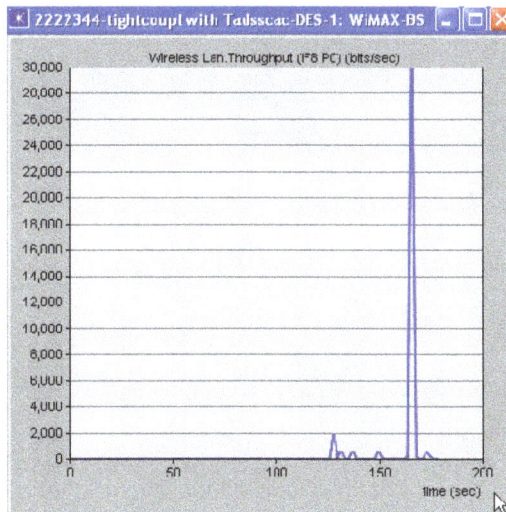

Fig.8 Throughput in WiMAX

Figs.8, 9 and 10 show the throughput, delay and load in WiMAX. Load created by users are less since three users are considered in this work that produces less throughput.

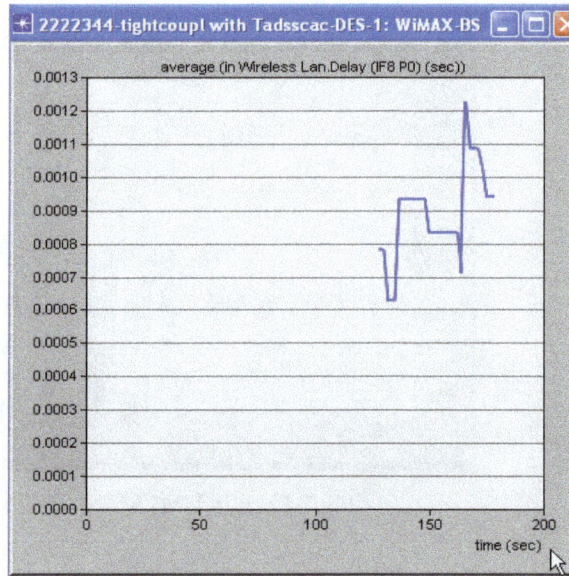

Fig.9 Delay in WiMAX

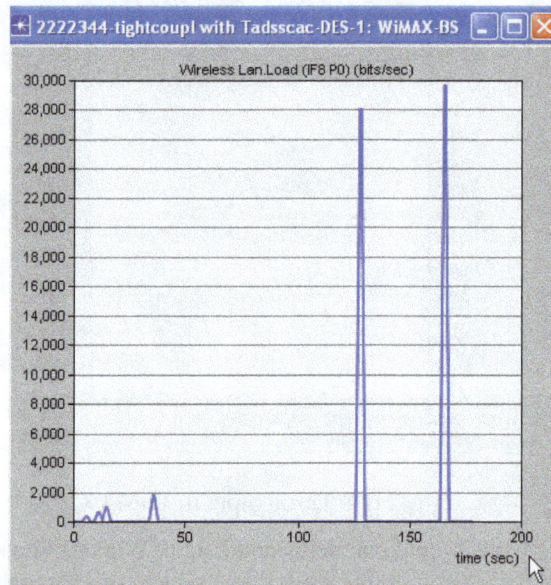

Fig.10 Load in WiMAX

Based on the metrics obtained in the integrated network architecture, network selection is performed. Initially user 1 (Voice) connected to WiMAX checks for the alternate networks (WLAN and UMTS) and calculates payoff in two networks. Fig.11 shows the payoff of user 1 in WLAN and UMTS. It is found that the payoff for user 1 in UMTS is maximum compared to WLAN. User 2 (Video) initially connected to WLAN checks for alternate networks and calculates payoff. The payoff for user 2 in WiMAX and UMTS is given in Fig.12. The figure shows that the payoff for user 2 in WiMAX is higher compared to UMTS. This is because user 1 already gets maximum payoff in UMTS.

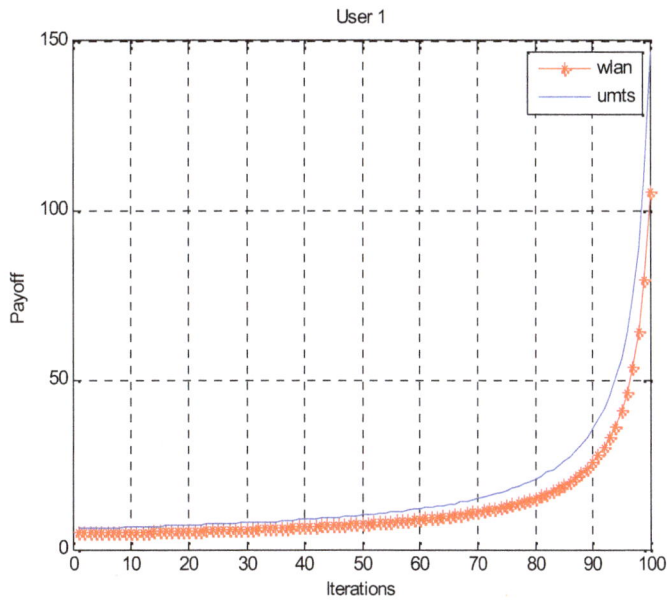

Fig.11 Payoff for User 1

User 3 (Web user) initially connected to UMTS checks for the alternate networks such as WLAN and WiMAX. Fig. 13 shows the payoff for user 3 in WLAN and WiMAX. The payoff for user 3 in WLAN is high when compared to WiMAX. This is because user 1 and 2 already gets maximum benefit in UMTS and WiMAX respectively.

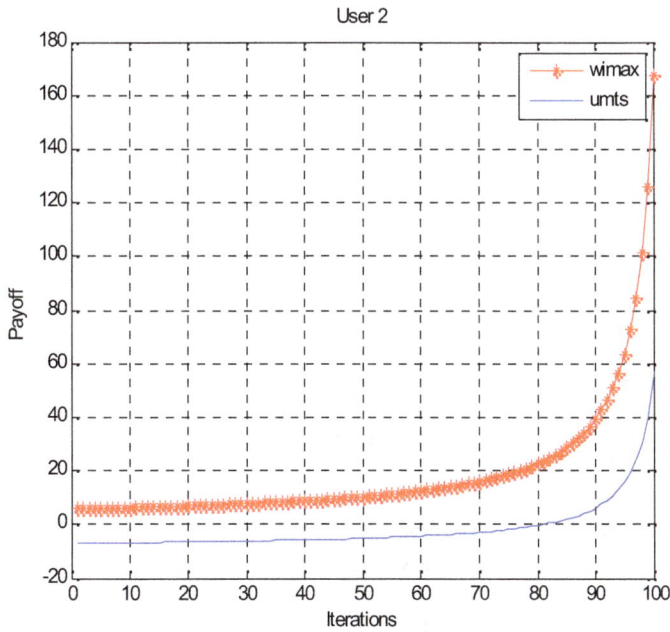

Fig.12 Payoff for User 2

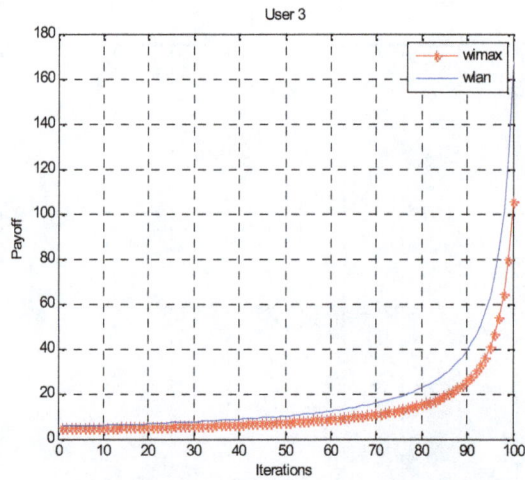

Fig.13 Payoff for User 3

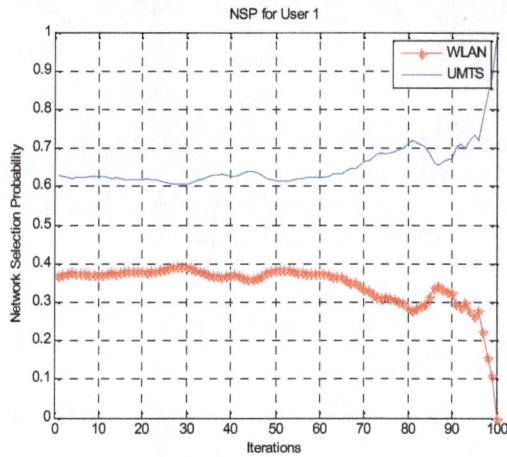

Fig.14 NSP for User 1

The NSP (Network Selection Probability) is determined based on the payoff as shown in Fig.14. Since the payoff for user 1 in UMTS is high, the NSP for user 1 in UMTS also high.Figs.15 and 16 show the NSP for user 2 and 3 respectively. Since payoff for user 2 in WiMAX is high so NSP for user 2 in WiMAX is also high. The maximum payoff for user 3 is attained in WLAN so NSP for user 3 in WLAN is high.

To implement the Quality of Experience (QoE) concept, users are classified as good and fair users. A good user is one who is ready to pay more for a particular service and expects high Quality of Service. On the other hand a fair user pays less for a service and compromises on the service quality. Fig.17 shows user 1 and user 3 are fair users and user 2 is a good user. Even though user 1 gets maximum payoff in UMTS since the user is fair user, so user 1 selects WLAN. User 2 is a good user and selects WiMAX. User 3 is a fair user already connected in WLAN and remains in WLAN.

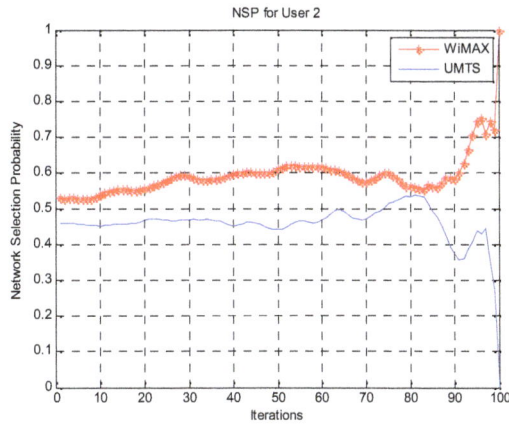

Fig.15 NSP for User 2

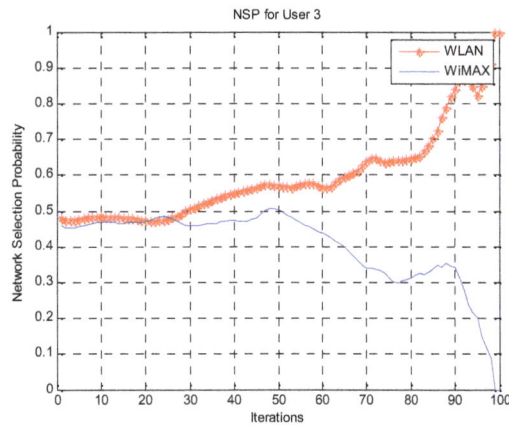

Fig.16 NSP for User 3

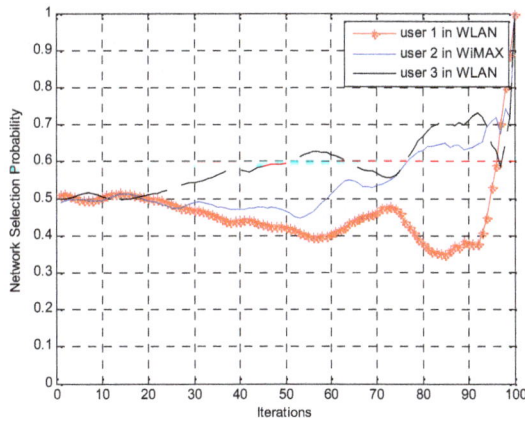

Fig.17 NSP for Users based on their budget

7. CONCLUSION

Heterogeneous networks require intelligent network selection process for seamless connectivity. In this paper WLAN, UMTS and WiMAX networks are integrated using IMS based on SIP. Network selection was performed using the proposed Boltzmann Gibbs based CODIPAS-RL because this algorithm performs network selection faster than the existing Bush Mosteller based CODIPAS-RL. i.e., Boltzmann Gibbs based CODIPAS-RL takes 90 iterations whereas Bush Mosteller based CODIPAS-RL requires 200 iterations for selecting a network. Three users with three different applications such as voice, video and web were considered and the network selection was performed. The payoff for each user in three different networks are determined and based on the payoff, network selection probability was obtained.

REFERENCES

[1]   D.E.Charilas and A.D. Panagopolous, "Multi access Radio Network Environments", In IEEE Vehicular Technology Magazine, vol. 5, no. 4, pp. 40-49, Dec. 2010

[2]   Q.T.Nguyen-Vuong, Y.Ghami-Doudane and N.Agoulmine, "On utility models for access network selection in wireless heterogeneous networks," *In IEEE Network Operations and Management Symposium (NOMS),* 2008.

[3]   H.Tembine and A.P.Azad, "Dynamic Routing Games: An Evolutionary Game Theoretic Approach," *IEEE conference on decision and control European Conference,* pp.4516-4521, Dec.2011.

[4]   M. Nam, N.Choi, Y.Seok and Y.Choi, "WISE: Energy-efficient interface selection on vertical handoff between 3G networks and WLANs," *Proceedings of IEEE International symposium PIMRC,* pp.692-698,2004.

[5]   Q.Y. Song and Abba S Jamalipour, " Network Selection in an Integrated Wireless LAN and UMTS environment using mathematical modelling and computing techniques", *IEEE Wireless Communications Magazine,* pp.42-48, June 2005.

[6]   Manzoor Ahmed Khan, Hamidou Tembine and Athanasis V. Vasilakos, " Game Dynamics and cost of Learning in Heterogeneous 4G Networks", *IEEE Journal on Selected Areas in Communications,* vol. 30, no. 1, pp.198-214, Jan. 2012.

[7]   Ramona Trestian, Olga Ormond and Gabriel-MiroMuntean, " Game Theory- Based Network Selection: Solutions and Challenges", *IEEE communications surveys and tutorials* ,pp.1-20, May 2011.

[8]   F.Zhu and J.McNair, "Optimizations for vertical handoff decision algorithms," *Proceedings of IEEE WCNC,* pp.867-872, 2004.

[9]   Mohammad SazidZaman Khan, ShaifulAlam and Mohammad RezaulHuque Khan, "A Network Selection Mechanism for Fourth Generation Communication Networks", *Journal of Advances in Information Technology,* vol. 1, no. 4, pp.189-196, Nov. 2010.

[10]   D.Niyato and E. Hossain, "Dynamics of network selection in heterogeneous wireless networks: An Evolutionary Game Approach," IEEE *Transactions on Vehicular Technology,* vol.58,no.4, pp.2008-2017, May 2009.

[11]   Samson Lasaulse, H.Tembine, Game theory and learning for wireless networks: Fundamentals and applications, *Elsevier,*2011.

[12]   P.Prabhavathi and L.Nithyanandan, "Network selection in wireless heterogeneous networks," *Proceedings of IEEE International Conference on Communication and Signal Processing – 2013,* pp. 1536-1539, April 2013.

[13]   D.Niyato and E. Hossain, "Dynamics of network selection in heterogeneous wireless networks: An Evolutionary Game Approach," IEEE *Transactions on Vehicular Technology,* vol.58,no.4, pp.2008-2017, May 2009.

[14]   H.Tembine, "Dynamic robust games in MIMO systems," *IEEE Transactions on systems Man and Cybernetics,* vol.99, no.41, pp.990-1002, Aug. 2011.

[15]   www.opnet.com

3

PAPR Reduction of OFDM signals using Selective Mapping with Turbo Codes

Pawan Sharma[1] and Seema Verma[2]

[1]Department of Electronics and Communication Engineering, Bhagwan Parshuram Institute of Technology, Rohini, New Delhi, India
Pawan061971@yahoo.co.in
[2]Department of Electronics and Communication Engineering, Banasthali University, Rajasthan, India
seemaverma3@yahoo.com

ABSTRACT

Multiple inputs multiple output orthogonal frequency division multiplexing (MIMO-OFDM) is an attractive transmission technique for high bit-rate communication systems. MIMO-OFDM has become a promising candidate for high performance 4G broadband wireless communications. One main disadvantage of OFDM is the high peak-to-average power ratio (PAPR) of the transmitter's output signal. Selected-Mapping (SLM) scheme which does not require the transmission of side information and can reduce the peak average power ratio (PAPR) in turbo coded orthogonal frequency division multiplexing (OFDM) system is proposed. Simulation results show that the system can achieve significant reduction in PAPR and satisfactory bit error rate performance over AWGN channels.

KEYWORDS

Orthogonal Frequency Division Multiplexing (OFDM), peak-to-average-power-ratio (PAPR), Selective Mapping (SLM), Complementary Cummulative Distribution Function (CCDF), Bit Error Rate (BER).

1. INTRODUCTION

Orthogonal frequency-division multiplexing (OFDM) has drawn significant interests over past decade [1], [2] for its robustness against the multipath fading channels. It is an effective high-speed data transmission scheme without using very expensive equalizers and it has been proposed as the air interface for broadband wireless applications such as wireless local area networks (WLANs). One of the major drawbacks of OFDM systems is that the OFDM signal exhibits a high peak-to-average power ratio (PAPR). Such a high PAPR necessitates the linear amplifier to have large dynamic range which is difficult to accommodate.

On the other hand, an amplifier with nonlinear characteristics will cause undesired distortion of the in-band and out-of-band signals. By now, many techniques have been proposed for relieving the PAPR problem in the OFDM, which can be roughly divided into two classes, the distortion-based techniques and the redundancy-based techniques. The distortion-based techniques reduce the PAPR of the OFDM symbol with the price of adding distortion to the signal points in the subcarriers. Direct clipping [1] simply suppresses the time-domain OFDM signals of which the signal powers exceed a certain threshold. The penalty is the significant increase of out-of-band energy. Peak windowing [2] or filtering after direct clipping [3] can be used to reduce the out-of-band energy. After the filtering operation, the peak of the time-domain signal may regrow. Hence, recursive clipping and filtering (RCF) [4] can be used to suppress both the out-of-band energy and the PAPR. RCF can be modified by restricting the region of distortion [5] to obtain improved error performance. On the other hand, estimation of the clipping noise at the receiver [6] can be used to improve the error performance of direct clipping or RCF.

The redundancy-based technique includes coding, selective mapping (SLM), partial transmit sequences (PTS), tone reservation (TR) and tone injection (TI) [7-13], etc. For the redundancy-based technique, the undesired effects occurring to the distortion based techniques can be alleviated while the penalty is the reduced transmission rate or increased average power due to the introduction of redundancy. A block coding technique [14] is to transmit only the code words with low PAPR. Such coding techniques offer good PAPR reduction performance and coding gain. Significant advance of the coding approach for PAPR control using generalized Reed-Muller codes is summarized in [15]. The critical problem for the coding approach is that for the OFDM system with large number of subcarriers, either it encounters design difficulties or the consequent coding rate becomes prohibitively low.

The basic idea of SLM technique is to generate several OFDM symbols as candidates and then select the one with the lowest PAPR for actual transmission. Conventionally, the transmission of side information is needed so that the receiver can use the side information to tell which candidate is selected in the transmission. In [16] and [17], the side information for a channel coded SLM appears explicitly in the data sequence to be encoded so that the side information is protected by the same channel code. The advantage of such an arrangement is that no additional protection is needed for side information and the rate loss due to the side information is small. However, once the side information is incorrectly decoded, the number of error bits in the erroneously decoded codeword can be great. In [12], an SLM technique (for either coded or uncoded cases) which does not need the transmission of side information was proposed, where the discrimination of the desired candidate against the undesired candidates is obtained by specially arranging the constellations for the subcarriers of each candidate so that the modulated signal points for the subcarriers of each pair of candidates are widely different.

In this paper we propose and examine a technique for reducing the probability of a high PAPR, based on part on a method proposed in [18] and [19]. This technique is a variation of selective mapping (SLM) [18], in which a set of independent sequences are generated by some means from the original signal, and then the sequence with the lowest PAPR is transmitted. To generate these sequences we use turbo encoder. Using turbo coding will offer two advantages, significant PAPR reduction and astonishing bit error rate (BER) performance.

2. PAPR PROBLEM AND SLM SCHEME

In the discrete time domain, an OFDM signal x_n of N subcarriers can be expressed as

$$x_n = \frac{1}{\sqrt{N}} \sum_{k=0}^{N-1} X_k \, e^{j2\pi kn/N} \qquad , \qquad 0 \le n \le N-1 \tag{1}$$

Where X_k, k = 0,1,2,3....., N-1 are input symbols modulated by BPSK, QPSK or QAM and n is the discrete time index.
The PAPR of an OFDM signal is defined as the ratio of the maximum to the average power of the signal, as follows

$$PAPR(x) = 10 \, log_{10} \frac{max\{|x_n|^2\}}{E\{|x|^2\}} \qquad 0 \le n \le N-1 \tag{2}$$

Where E{.} denotes the expected value operation and $x = [x_1, x_2, x_3, \ldots \ldots x_{N-1}]^T$

As to the discrete-time signals, since symbol-spaced sampling may sometimes miss some of the signal peaks, signal samples are obtained by oversampling by a factor of L to better approximate the true PAPR. Oversampled time-domain samples are usually obtained by LN-point IFFT of the data block with $(L-1) N$ zero-padding. It is shown in [20] that $L = 4$ is sufficient to capture the peaks.

When the OFDM signal with high PAPR passes through a non-linear device, (power amplifier working in the saturation region), the signal will suffer significant non-linear distortion [20]. This non-linear distortion will result in in-band distortion and out-of-band radiation. The in-band distortion causes system performance degradation and the out-of-band radiation causes adjacent channel interference (ACI) that affects systems working in the neighbour bands. To lessen the signal distortion, it requires a linear power amplifier with large dynamic range. However, this linear power amplifier has poor efficiency and is so expensive. Obviously, the distribution of PAPR bears stochastic characteristics in a practical OFDM system. Usually, Complementary Cumulative Distribution Function (CCDF) can be used to evaluate the performance of any PAPR reduction schemes, given by [17].

$$CCDF(N, PAPR_0) = 1 - (1 - e^{-PAPR_0})^N \qquad (3)$$

In the SLM approach, U statistically-independent phase sequences, say, $P^{(u)} = [P_0^{(u)}, P_1^{(u)}, P_2^{(u)}, \ldots \ldots P_{N-1}^{(u)}]^T$ are generated, where $P_k^{(u)} = \exp(j\Phi_k^{(u)})$, $\Phi_k^{(u)} \in [0, 2\pi]$, , $k = 0,1,2,\ldots\ldots,$ N-1, u $= 1,2,3,\ldots\ldots,$ U. Then the data block $X = [X_0, X_1, X_2, \ldots \ldots X_{N-1}]^T$ is multiplied component-wise with each one of U different phase sequence $P^{(u)}$, resulting in a set of U different data blocks $X^u = [X_0 P_0^{(u)}, X_1 P_1^{(u)}, X_2 P_2^{(u)}, X_3 P_3^{(u)}, \ldots X_{N-1} P_{N-1}^{(u)}]^T$, $u = 1,2,3,\ldots,$ U . Then, all U alternative data blocks (one of the alternative subcarrier sequences must be the unchanged original one) are transformed into time domain to get transmitted symbols x^u , u = 1,2,3,......,U by IFFT, where x^u, u = 1,2,3,......,U are defined as the candidate signals. Finally, the one with the minimum PAPR is selected for transmitting, shown in Fig. 1.

At the side of receiver, in order to recover the received signals successfully, the side information is required. This information must be transmitted accompanying with the transmitted signal. When binary symbols are used, $[log_2 U]$ bits are required to represent this side information [2, 3], where operation [.] rounds the elements to the nearest integers toward infinity.

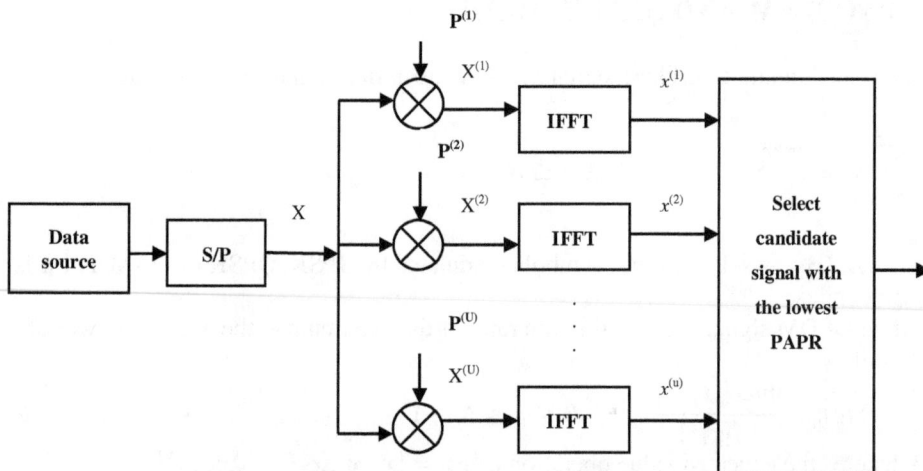

Figure 1. The block diagram of SLM scheme

2.1. Selective mapping using Turbo Coding

The probability that, the PAPR of the OFDM signal exceeds a certain threshold Y is given by

$$P_r\{PAPR > Y\} = 1 - (1 - e^{-Y})^N \tag{4}$$

In SLM it is assumed that, U statistically independent alternative sequences, which represent the same information, are generated by some suitable means. The sequence with the lowest PAPR is selected for transmission.

The probability that, the lowest PAPR Y_l exceeds a certain threshold Y *is* given by

$$P_r\{Y_l > Y\} = (P_r\{PAPR > Y\})^U \tag{5}$$

To generate these sequences linear feedback shift register (LFSR) is used [11]. A LFSR is used to transform the data before it is mapped to the orthogonal channels. Different sequences are generated by inserting different bits labels at the beginning of the data. This results in $U = 2^m$ different sequences, where m is the length of the inserted bits.

Turbo codes [21] are parallel concatenated convolutional codes in which the information bits are first encoded by a recursive systematic convolutional (RSC) code and then, after passing the information bits through an interleaver, are encoded by a second RSC code. Turbo decoder is used to recover the transmitted signal at the receiver side. The Turbo decoder consists of two soft input soft output (SISO) modules [22], an interleaver and de-interleaver. Figure 2 shows a turbo system, turbo encoder and decoder.

In this paper, instead of using LFSR, we use turbo encoder to generate different sequences and the sequence with the lowest PAPR is selected for transmission. The different sequences are generated by inserting different bits labels at the beginning of the data. Figure 3 shows the transmitter side of an OFDM system, where the turbo coding and SLM are used for PAPR reduction. For each bits labels b_i, i = 1, 2, 3 ...., U where b_i, a sequence of m bits, the turbo encoder will generate a sequence xi, i = 1, 2, ..., U. The sequence that has the lowest PAPR will be selected for transmission. At the receiver side, the receiver does not need any side information, and the bits labels are discarded after decoding.

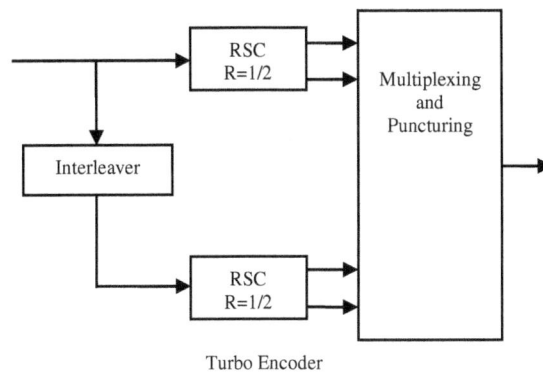

Turbo Encoder

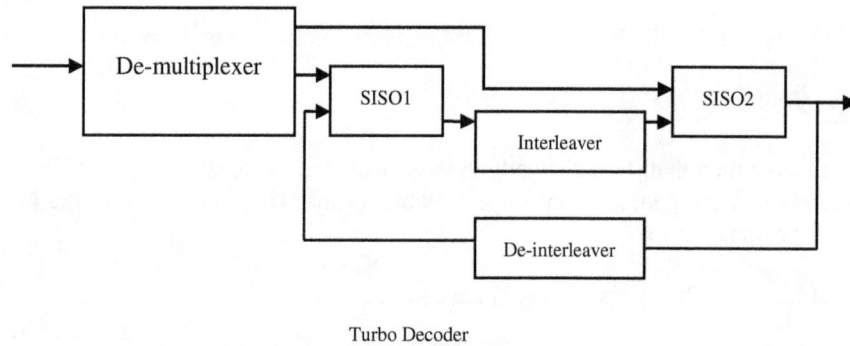

Figure 2. Turbo System

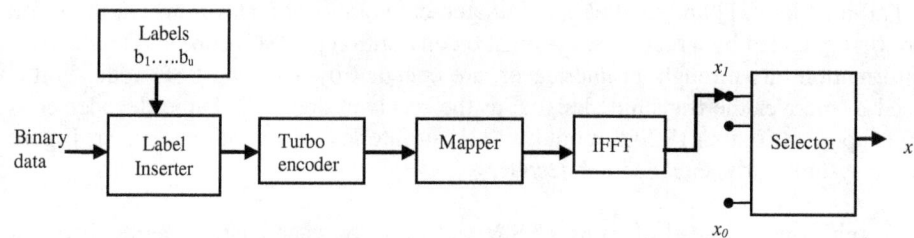

Figure 3. Turbo Coded OFDM System Model

2.2. Simulation Results

The PAPR reduction and BER performances of the proposed scheme are examined by computer simulation. In the simulation we consider an OFDM signal with $N = 128$ subcarriers, 16-Quadrature Amplitude Modulation (16-QAM) mapping, and turbo code with two RSC encoders each has a constraint length $K = 4$ with generator polynomial 15/17, where $g1/g2$ represents the forward/backward generator polynomials in octal base. Puncturing is used to increase the overall code rate to $R = 1/2$. To obtain accurate PAPR calculations the signal is over sampled, $L = 8$. Additive white Gaussian noise (AWGN) channel is assumed. At the receiver side logarithmic maximum a posterior probability (log-MAP) algorithm is used to implement the SISO modules.

Figure 4.shows the complementary cumulative distribution function (CCDF) of the PAPR of the OFDM signal, where turbo coding and SLM are used for PAPR reduction. The CCDF of the PAPR without SLM is also shown (U=1). Clearly, the probability of high PAPR is reduced significantly as the number of sequences increases. When $U = 1$, the PAPR is approximately 9.7 dB and with turbo coding and selective mapping, with $U =32$, the PAPR of the OFDM signal is nearly 6.8 dB. The simulation results agree with the results obtained by the approximation in equation (4) with small differences. These small differences due to the fact that, equation (4) was derived with the assumption that, the samples are mutually uncorrelated which is not true anymore when over sampling is applied. The BER performance of the proposed scheme is also shown in Figure 5. For number of iterations 1, 2 and 5, the probability of bit error is reduced upto $8 \times 10-5$.

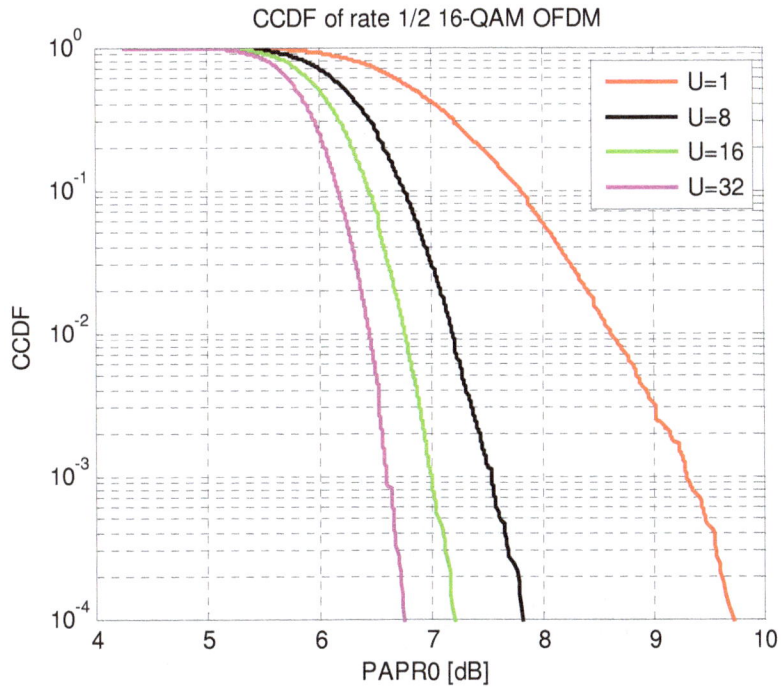

Figure 4. PAPR Performance with 16 QAM

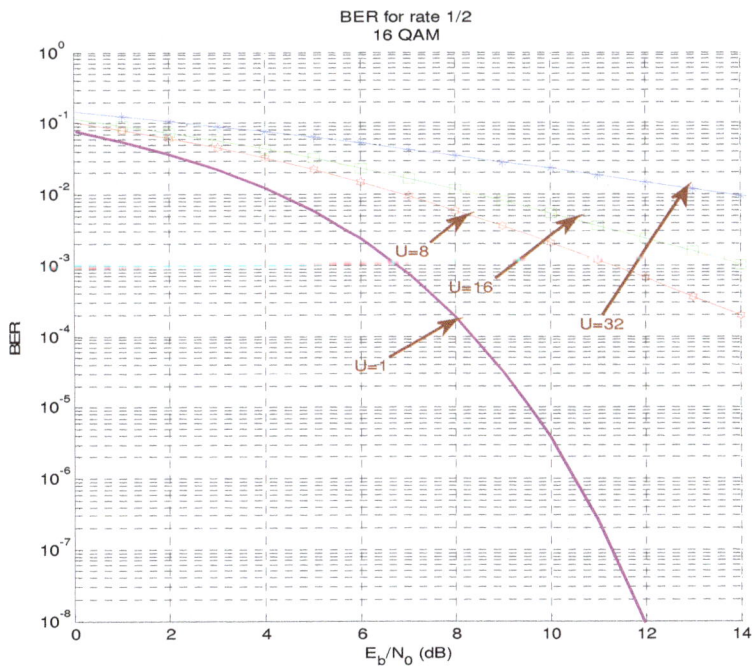

Figure 5. Bit Error Rate (BER) Vs. SNR

2.3. CONCLUSIONS

We have shown that, Turbo coding and SLM can be combined to reduce the PAPR of OFDM signal with quite moderate additional complexity. The advantage of the proposed scheme is that, the Turbo encoder is used for two purposes, error correction and PAPR reduction. This reduces the hardware complexity of the system.

REFERENCES

1. S.H.Han and J.H.Lee,"An Overview of Peak-to-Average Power Ratio Reduction Techniques For Multicarrier Transmission," IEEE Wireless Communications,Vol.12,No.2, Apr.2005, pp.56-65.

2. T.Jiang and Y.Wu, "An Overview: Peak-to-Average Power Ratio Reduction Techniques for OFDM Signals," IEEE Transactions on Broadcasting, Vol.54, No.2, June 2008, pp.257-268.

3. X.Li and L.J.Cimini," Effects of Clipping and Filtering on the Performance of OFDM," IEEE Communication Letters, Vol.2, No.5, May 1998, pp.131-133.

4. J.Armstrong, "Peak-to-Average Reduction for OFDM by Repeated Clipping and Frequency Domain Filtering," IEEE Electronics Letters, Vol.38, No.5, May 2002, pp.246-247.

5. K.D.Rao and T.S.N.Murthy, "Analysis of Effects of Clipping and Filtering on the Performance of MB-OFDM UWB Signals," Proc. of the 2007 15th International Conference on Digital Signal Processing (DSP 2007), IEEE, pp.559-562.

6. H.Ochiai and H.Imai, "Performance Analysis of deliberately clipped OFDM signals," IEEE Trans. on communications, Vol.50, No.1, January 2002, pp.89-101.

7. G.Hill and M.Faulkner, "Comparison of Low Complexity Clipping Algorithms for OFDM," IEEE International Symposium on Personal, Indoor and Mobile Radio communications, 2002,Vol..1, pp.227-231.

8. S.K.Yusof and N.Fisal," Coorelative Coding with Clipping and Filtering Technique in OFDM Systems," ICICS-PCM 2003,Singapore, IEEE 2003, pp.1456-1459.

9. A.Ghassemi and T.A. Gulliver, "PAPR Application of OFDM Using PTS and Error-Correcting Code Subblocking," IEEE Trans. On Wireless Communications, Vol.9, No.3, March 2010, pp.980-989.

10. S.H.Han and J.H.Lee, "PAPR Reduction of OFDM Signals using a Reduced Complexity PTS Technique," IEEE Signal Processing Letters, Vol.11, No.11, Nov.2004, pp.887-890.

11. H.Breiling,S.H.Muller-Weinfurtner,and J.B.Huber, "SLM Peak-Power   Reduction without Explicit Side Information," IEEE Communication Letters, Vol.5, No.5, June 2001, pp.239-241.

12. A.D.S.Jayalath and C.Tellambura, "SLM and PTS Peak-Power Reduction of OFDM Signals without side information," IEEE Transactions on Wireless Communications, Vol.4, Sept.2005, pp.2006-2013.

13. S.Y.Le Goff, Al-Samahi, S.S.B.K.Khoo, C.C.Tsimenidis and B.S.Sharif, "Selected Mapping without side information for PAPR Reduction in OFDM," IEEE Transactions on Wireless Communications, Vol.8, No.7, July. 2009, pp.3320-3325.

14. Z.Q.Taha and X.Liu,"An Adaptive Coding Technique for PAPR Reduction," Global Telecommunication conference, Globecom 2007, IEEE , pp.376-380.

15 K. Paterson, "Generalized Reed-Muller codes and power control in OFDM modulation," *IEEE Trans. Inform. Theory*, vol. 46, no. 1, pp. 104–120, Jan. 2000.

16 H. Breiling, S.H.Muller-Weinfurtner, and J.B.Huber, "SLM Peak-Power Reduction without Explicit Side Information," IEEE Communication Letters, Vol.5, No.5, June 2001, pp.239-241.

17 H.Ochiai and H. Imai, "On the distribution of the peak-to-average power ratio in OFDM signals," *IEEE Trans. Commun.*, vol. 49, pp. 282–289, Feb. 2001.

18   R. Fischer and J. Huber,"Reducing the peak to average power ratio of multicarrier modulation by selected mapping," Elect. Letts. vol. 32, pp. 2056-2057, Oct. 1996.

19 N. Carson, T. A. Gulliver, "PAPR reduction of OFDM using selected mapping, modified RA codes and clipping", in Proc. IEEE VTC, vol.2, pp. 1070- 1073, Sep. 2002.

20 C. Tellambura, "Computation of the continuous-time PAR of an OFDM signal with BPSK subcarriers", IEEE Commun. Letter. vol. 5, no. 5, May 2001, pp. 185-187.

21 C. Berrou, A. Glavieux and P. Thitimajshima,"Near Shannon limit error correcting coding and decoding: Turbo Codes." in Proc. Of ICC'93, (Geneva, Switzerland), pp. 1064-1070, May 1993.

22 S. Benedetto, G. Montorsi, D. Divsalar, and F. Pollara, "A soft-input soft-output maximum a posteriori (MAP) module to decode parallel and serial concatenated codes", TDA progress report 42-127, Jet Propulsion Lab., Pasadena, CA, Nov. 15 1996.

4

A Pixel Domain Video Coding based on Turbo code and Arithmetic code

Cyrine Lahsini[1], Sonia Zaibi[2], Ramesh pyndiah[1] and Ammar Bouallegue[2]

[1] Signal and Communication Department, Telecom Bretagne, France
Email: `firstname.name@enst-bretagne.fr`
[2] Syscoms Laboratory, National Engineering School of Tunis, Tunisia
Email: `firstname.name@enit.rnu.tn`

Abstract

In recent years, with emerging applications such as multimedia sensors networks, wireless low-power surveillance and mobile camera phones, the traditional video coding architecture in being challenged. In fact, these applications have different requirements than those of the broadcast video delivery systems: a low power consumption at the encoder side is essential.

In this context, we propose a pixel-domain video coding scheme which fits well in these senarios. In this system, both the arithmetic and turbo codes are used to encode the video sequence's frames. Simulations results show significant gains over Pixel-domain Wyner-Ziv video codeingr.

Keywords : Turbo code, arithmetic code, distributed video coding.

1 Introduction

In conventional video coding such as $H26x$, the complexity of encoder is much higher than that of decoder. However, in many applications such as sensor networks and multi-cameras scenarios where low-power and low-complexity encoder device is essential, new types of coding algorithms have to be proposed.

Distributed video coding (DVC), based on the Slepian-Wolf and Wyner-Ziv theorems ([2],[1]), fits well these emerging scenarios since it enables the exploitation of the video statistics, partially or totally, at the decoder only. A flexible allocation of complexity between the encoder and the decoder is therefore enabled by the DVC paradigm. Its core parts is a Slepian-Wolf encoder which often involves turbo codes because of their strong error correction capabilities.

The turbo encoder generates parity bits from the Wyner-Ziv frames (WZ frames) which are sent to refine the side information constructed at the decoder by frame interpolation using the neighboring key frames (K frames) already received.

The rate distorsion and complexity performance of distributed video codec depend on the group of picture (GOP) size i.e. the number of wyner-ziv frames between two key frames. It is important to notice that the larger the GOP size, the lower is the overall complexity since the WZ frame encoding process is less complex that the K frame encoding one.

To achieve low complexity encoding, we propose in this paper a pixel-domain video coding scheme in which both arithmetic and turbo codes are used to encode the video sequence's frames.

We consider a GOP size greater that one, the key frames (as in the distributed video coding) are encoded and decoded using an intra codec. For the GOP remaining frames, we exploit the temporal correlation using an entropy encoder (artithmetic encoder) for only the two most significant bitplanes. The other bitplanes are encoded using a turbo code.

Simulations results obtained with our proposed framework show significant gains comparing Pixel-domain Wyner-Ziv video codec.

The paper is organised as follows, in section II, we describe the distributed video coding scheme in pixel domain. In section III, we present our proposed scheme in which both arithmetic and turbo codes are used. In section IV, we discuss the simulation details and compare the performance of the proposed coder to the pixel-domain Wyner-Ziv coding. Finally, section V concludes this paper.

2   Pixel Domain Distributed Video Coding

According to Slepian and Wolf, it is possible to encode separately and decode jointly two statistically dependent signals X and Y. They show, in [1], that the possible rate combinations of R_X and R_Y for a reconstruction of X and Y with an arbitrarily small error probability are expressed by :

$$R_X \geq H(X \mid Y)$$

$$R_Y \geq H(Y \mid X)$$

$$R_X + R_Y \geq H(X,Y)$$

where $H(X,Y)$ is the joint entropy of X and Y, $H(X \mid Y)$ and $H(Y \mid X)$ are their conditional entropies.

Figure 1 illustrates the achievable rate region for which the distributed compression of two statistically dependent sources X and Y allows recovery with an arbitrarily small error probability.

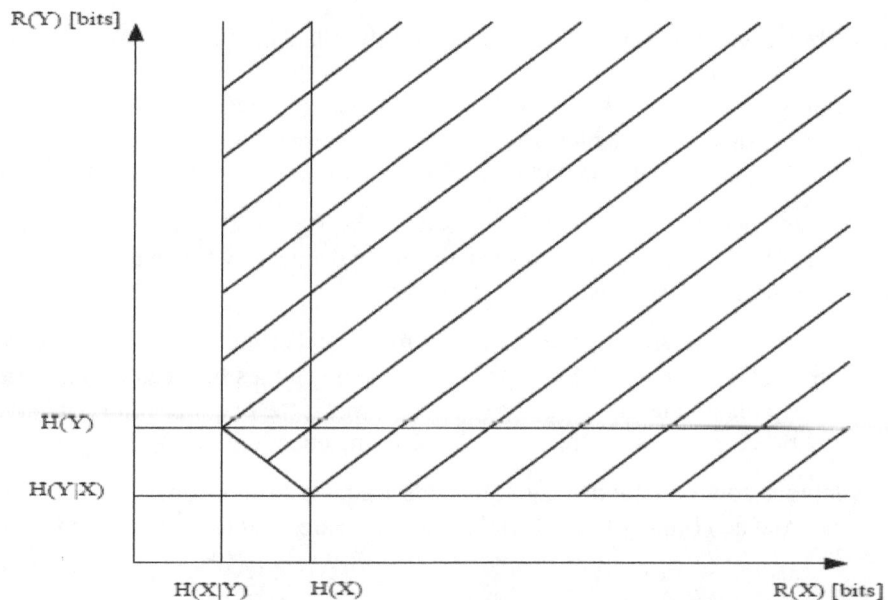

Figure   1:   Achievable rate region following the Slepian-Wolf theorem

Wyner and Ziv have studied a particular case of Slepian-Wolf coding corresponding to the rate point $(H(X|Y), H(Y))$. This particular case deals with the source coding of the X sequence considering the Y sequence, known as side information, is available at the decoder ([2],[3]).

In general, a Wyner-Ziv encoder is comprised of a quantizer followed by a Slepian-Wolf encoder. The Wyner-Ziv coding paradigm may be applied in the pixel domain or in the transform domain. The pixel domain coding solution is the most simplest in terms of complexity.

We introduce in this section the basic pixel domain DVC for a GOP equal to 1 (one WZ frame between two K frames). It is shown in Figure 2 and it follows the same architecture as the one proposed in [5].

Let $X_1, X_2 \ X_N$ be the frames of a video sequence. The odd frames, X_{2i+1} where $i \in \{0,1,..(N-1)/2\}$, are the key frames (K frames) which are available at the decoder. Each even frame (Wyner Ziv frame WZ) X_{2i}, where $i \in \{0,1,..N/2\}$, is encoded independently of the key frames and the other even frames.

X_{2i} is encoded as follows:

- We scan the frame row by row and quantize each pixel using an uniform 2^M levels quantizer, generating the quantized symbol stream q_{2i}.

- Over the resulting quantized symbol stream (constituted by all the quantized symbols of X_{2i} using 2^M levels bitplane extraction is performed and each bitplane is then independently turbo encoded.

- The redundant (parity) information p_{2i} produced by the turbo encoder for each bitplane is stored in a buffer and transmitted in small amounts upon decoder request via the feedback channel.

- For each frame X_{2i}, the decoder takes the adjacent key frames X_{2i-1} and X_{2i+1} and performs temporal interpolation $Y_{2i} = I(X_{2i-1}, X_{2i+1})$. The turbo decoder uses the side information Y_{2i} and the received subset of p_{2i} in the bitplanes decoding by computing the Log Likelihood Ratio.

- The side information is, also, used in the reconstruction process.

- To make use of the side information Y_{2i}, the decoder needs some model for the statistical dependency between X_{2i} and Y_{2i}. The correlation between the original frame X_{2i} and the side information Y_{2i} is described by a noise model where Y_{2i} is a noisy version of the original frame X_{2i}. The statistical model is necessary for the conditional probability calculations in the turbo decoder.

What it is done in the literature (see [3],[5]) is to consider that the virtual noise between X_{2i} and Y_{2i} has a laplacian distribution with zero mean and estimated standard deviation $1/\alpha$.

For more details of the used DVC scheme, see [3],[5].

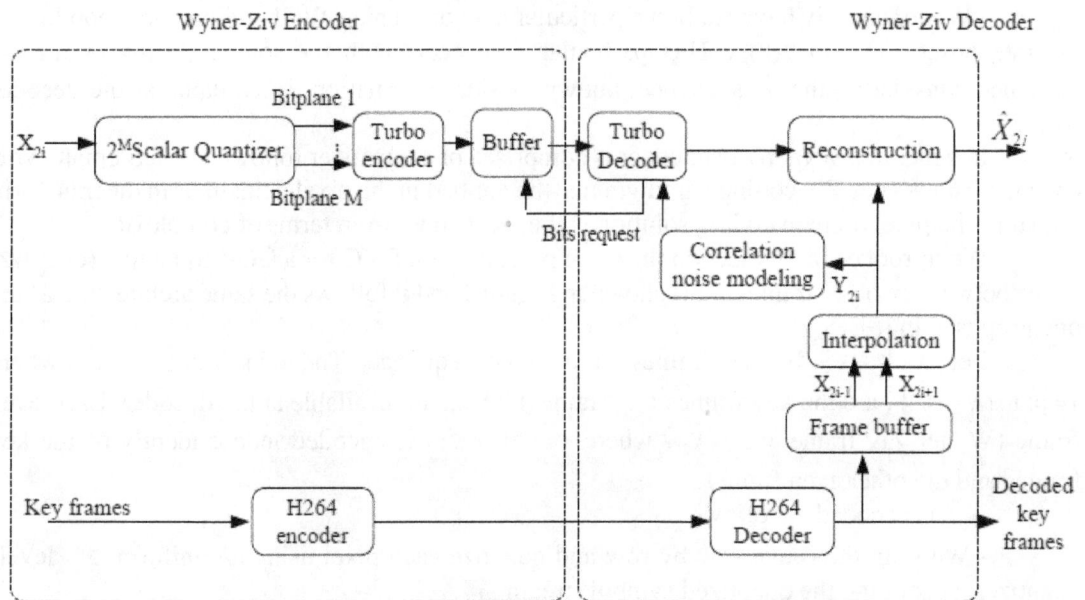

Figure 2: Distributed video CoDec in pixel Domain

The same principle is applied for greater sizes of GOP. We present, in figure 3, the pixel domain DVC for a GOP size equal to 3. The Wyner Ziv frames are noted $W1$, $W2$ and $W3$ and the key frames are noted $K1$ and $K2$. This scheme will be the reference for our proposed framework.

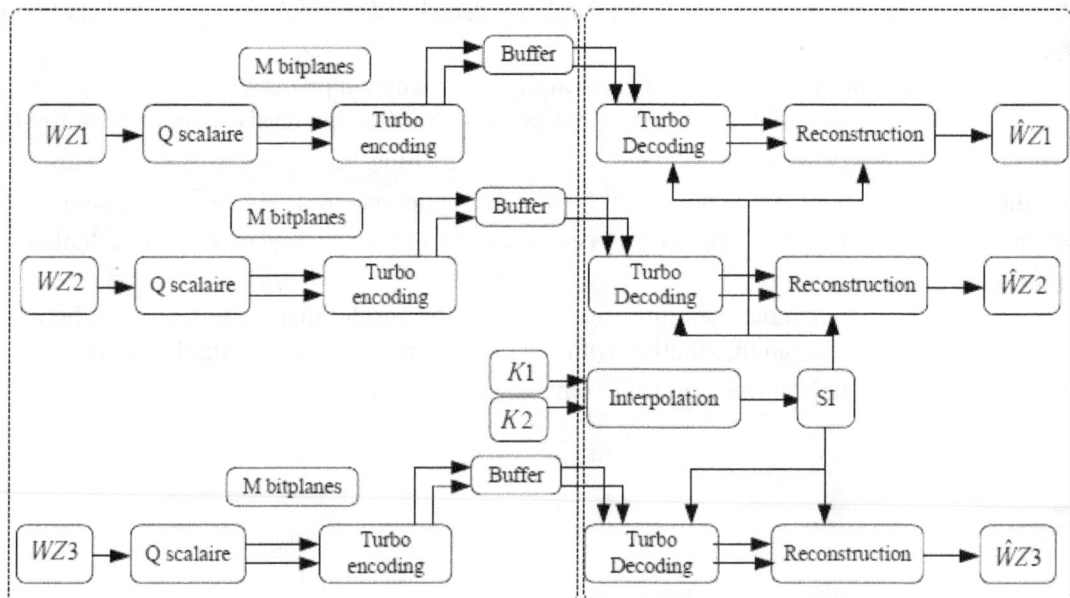

Figure 3: Distributed video CoDec in pixel Domain for a GOP size =3

3 A low complexity video coding scheme based on turbo code and arithmetic code

In order to improve the rate/distortion performance of video coding scheme with low complexity at the encoder, we reduce the frequency of key frames K by increasing the size of the GOP(Group Of Picture). Indeed, the encoding of key frames in intra mode requires a high rate compared to the remaining frames encoding pictures (called F frames in this section).

We propose a new architecture of video coding scheme that integrates both the turbo code and the arithmetic code. The proposed framework presents better performances than the distributed video coding scheme considered in Section 2.

Consider a GOP size = 3 (Figure 4, the algorithm can be extended to larger GOP size. In this section we present the procedure of GOP's frames encoding/decoding.

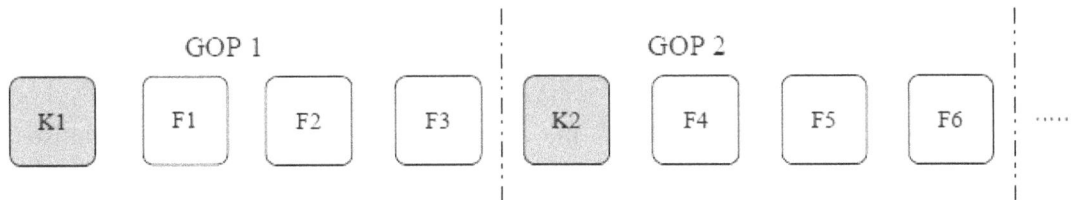

Figure   4: GOP=3

3.1 Proposed framework

It is known that the frames in a video sequence are characterized by a high correlation in both the spatial and temporal domain. Our scheme is based on the exploitation of the temporal correlation between the successives frames. If we note by $K1$, $F1$, $F2$ and $F3$ the frames of the GOP, the corresponding most significant bitplanes (the two most significant bitplanes) are highly correlated temporally. In other words, the binary error between two successive bitplanes (of two successive frames) is low. Thus, the application of an entropy coder provides a good compression. The frames will be processed successively.

Our algorithm follows these steps:

- The key frame $K1$ is encoded by an intra encoder and the bitstream is transmitted to the decoder.

- The encoder performs the intra decoding of the bitstream (generated by the intra encoding of $K1$) and the resulting frame is the reference frame noted (REF) and used to encode the first frame $F1$.

- We use a scalar quantizer with M levels (we consider $M = 4$) to quantify both the REF and $F1$ frames.

- Over the resulting quantized symbol streams X and Y (constituted respectively by all the quantized symbols of REF and $F1$ frames using 2^M levels) bitplane extraction is performed to generate M bitplanes for each frame:

$X = (X_1, X_2, ...X_M) = (X_1, X_2, X_3, X_4)$ Bitplane extraction (from Most Significant Bitplane MSB to Least Significant Bitplane LSB) of the REF frame.

$Y = (Y_1, Y_2, ...Y_M) = (Y_1, Y_2, Y_3, Y_4)$ Bitplane extraction (from Most Significant Bitplane MSB to Least Significant Bitplane LSB) of the $F1$ frame.

Least significants bitplanes encoding : Frame F1

The least significant bitplanes Y_3 and Y_4 are encoded separately using a turbo encoder according to the scheme shown in Figure 2. The blocks of reconstruction and interpolation will not be considered at this level.

The turbo encoder produces a sequence of parity bits (redundant bits related to the initial data) for each bitplane array; the parity bits are then stored in a buffer, punctured, according to a given puncturing pattern, and transmitted upon decoder request via the feedback channel.

Least significants bitplanes decoding : Frame F1

- The decoder performs frame interpolation using the previous and the next temporally adjacent frames of F_k, $k = 1..3$ to generate the frame Y_{2i} an estimate of F_k.

- A bitplane extraction is then carried out over the interpolated frame Y_{2i}. The residual statistics between correspondant pixels in F_k and Y_{2i} is assumed to be modelled by a Laplacian distribution. The laplacian parameter is estimated offline at frame level.

- Once Y_{2i} and the residual statistics are known, the decoded quantized symbol stream associated to the frame F_k is obtained through an iterative turbo decoding procedure.

• Channel noise model

To use the side information Y_{2i}, the decoder needs some model for the statistical dependency between F_k and Y_{2i}. The statistical model is necessary for the conditional probability calculations in the turbo decoder.

What it is done in the litterature (see [3], [5]) is to consider that the virtual noise between F_k and Y_{2i} has a laplacian distribution with zero mean and estimated standard deviation $\frac{1}{\alpha}$

For every possible value of the pixel with amplitude x, the probability that $F_k(r,c)$ is equal to x is evaluated as :

$$p[F_k(r,c) = x] = \frac{\alpha}{2} \exp(-\alpha|x - Y_{2i}(r,c)|) \tag{1}$$

Let then $F_k(r,c)$ be the jth bit of the value F_k and let Z_j be the set of x values that have j^{th} bit equal to zero.

Then, for every j we compute :

$$p[F_k(r,c) = 0] = \sum_{x \in Z_j} p[F_k(r,c) = x] \tag{2}$$

• Turbo decoder

The turbo code used in this paper consists of two recursive systematic convolutional encoders. The decoding procedure of turbo code is performed by using a modified version of the Maximum A Posteriori (MAP) algorithm ([7],[8]).

For the MAP decoder, the soft decision of each transmitted bit i_t (t is the time index) is given by :

$$L(\hat{i}_t) = ln(\frac{P(i_t = +1|o)}{P(i_t = -1|o)}) \tag{3}$$

where $P(i_t = +1|o)$ is the a posteriori probability.

The symbol $O = (o_0, o1, ..., o_t, ..., o_{N-1})$ with $o_t = (o_t^s, o_t^p) = (Y_t, P_{it})$ represents the information at the turbo decoder for each i_t : that is, the parity bit P_{it} (i represents the index of the RSC encoder from which P_t belongs to) and the systematic bit Y_t.

Incorporating the convolutional code's treillis, Eq. 3 may be written as :

$$L(\hat{i}_t) = L_c + L_{12}^e(i_t) + L_{21}^e(i_t) \tag{4}$$

for MAP decoder 2. In Eq. 4, $L_c = ln(\frac{P(o_t^s | i_t = +1)}{P(o_t^s | i_t = -1)})$ is the information determined from the laplacian virtuel channel. $L_{21}^e(i_t)$ represents the extrinsic information generated by MAP decoder 2 and is to be used as *apriori* information by MAP decoder 1. $L_{12}^e(i_t)$ is the *apriori* information generated by MAP decoder 1.

To calculate $L(\hat{i}_t)$, we need to determine two variables : $\tilde{\alpha}_t$ and $\tilde{\beta}_t$ *(for more details, refer to [7])*.

The probability $\tilde{\alpha}_t(S_t)$ can be computed from :

$$\tilde{\alpha}_t(S_t) = \frac{\sum_{S_{t-1}} \tilde{\alpha}_{t-1}(S_{t-1}) \times \gamma_t(S_{t-1}, S_t)}{\sum_{S_t}\sum_{S_{t-1}} \tilde{\alpha}_{t-1}(S_{t-1}) \times \gamma_t(S_{t-1}, S_t)} \tag{5}$$

where S_t is a state of the RSC encoder. Considering that the initial state of the RSC encoder, S_0, is known, the initialization of $\tilde{\alpha}_t(S_t)$ (i.e for $t = 0$) is given by :

$\tilde{\alpha}_0(S_t) = 1, S_t = S_0$
$\tilde{\alpha}_0(S_t) = 0, S_t \neq S_0$

The probability $\tilde{\beta}_{t-1}(S_{t-1})$ can be computed from :

$$\tilde{\beta}_{t-1}(S_{t-1}) = \frac{\sum_{S_t} \tilde{\beta}_t(S_t) \times \gamma_t(S_{t-1}, S_t)}{\sum_{S_t}\sum_{S_{t-1}} \tilde{\alpha}_{t-1}(S_{t-1}) \times \gamma_t(S_{t-1}, S_t)} \tag{6}$$

is a state of the RSC encoder.

Considering that the final state of the RSC encoder, S_N, is known :

$\tilde{\beta}_N(S_t) = 1, S_t = S_N$
$\tilde{\beta}_N(S_t) = 0, S_t \neq S_N$

The probability $\gamma_t(S_{t-1}, S_t)$ can be written as :

$$\gamma_t(S_{t-1}, S_t) = P(i_t)p(o_t | i_t) \tag{7}$$

where $p(o_t | i_t) = p(o_t^s | i_t)p(o_t^p | i_t)$ since the parity information "error" is independent of the systematic information "error".

As was already mentioned before, the side information Y provides the systematic information to the turbo decoder. Then, $p(o_t^s | i_t) = p(Y_t | i_t)$ is modelled by the Laplacian distribution given by Eq.2.

In this paper, it is assumed that no errors are introduced in the parity bits transmission. The conditional pdf $p(o_t^p | i_t)$ may be descibed by a gaussian distribution with mean zero and a variance σ^2 arbitrarily small.

Most significants bitplanes encoding : Frame F1

For the most significant bitplanes Y_1 and Y_2, we calculate the vector $Z = \{Z_1, Z_2\}$ where

$$Z_i = X_i \oplus Y_i$$

$\oplus$ is the sum modulo-2 performed bit by bit according to the weight.

The binary vectors Z_1 and Z_2 are representing the bits that have changed between the reference frame *REF* and the frame *F1* i.e. the error between the binary bitplanes (the two most significant ones).

Since the video sequence is characterized by high temporal correlation, the binary error is, also, a sequence of highly correlated bits. To exploit this characteristic, the vectors Z_1 and Z_2 are encoded independently by an entropy encoder (arithmetic encoder) to form the bit stream that is transmitted to the decoder.

Most significants bitplanes decoding : Frame F1

At the reception, the decoder decodes the received bit stream using an arithmetic decoder. We obtain the binary sequence $\tilde{Z}_i$, $i=1,2$ Using the the *REF* frame (available at the decoder), we reconstruct the most significant bitplanes of the decoded frame *F1* :

$$\tilde{Y}_i = \tilde{Z}_i \oplus \tilde{X}_i$$

where $\tilde{X}_i = \tilde{X}_1, \tilde{X}_2$ are the decoded bitplanes of the *REF* frame using the intra decoder.

The process of encoding/decoding of the most significant bitplanes is represented by Figure 5.

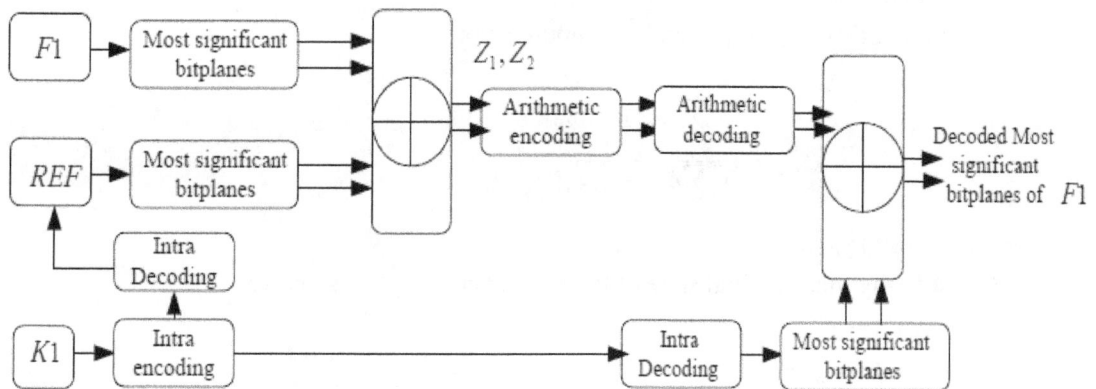

Figure   5: Most significant bitplanes encoding/decoding

Encoding/Decoding :frames F2 and F3

For frames *F2* anf *F3* , the same procedure described for the coding of the *F1* frame is used, using respectively *F1* and *F2* as *REF* frame.

Reconstruction of frames F1, F2 et F3 :

The reconstruction module makes use of the decoded quantized symbol stream and Y_{2i} to reconstruct each pixel of the *F1*, *F2* and *F3* frame.

3.2   Proposed framework : improvements

• Decimal level Arithmetic Encoder/Decoder

To better improve performance of DVC codec, we have modified the scheme proposed above (Figure 5). We maintain the same steps outlined previously in Section 3.1. Only the most significant bitplanes processing is modified as follows :

Let us recall that the binary vectors $X_1, X_2, Y_1 and Y_2$ are defined by:

$X_1 = \left(x_{1_1}, x_{1_2}, x_{1_N}\right)$, where $\left\{x_{1_i}\right\}$, $1 \le i \le N$ represent the most significant bits obtained after quantification of the REF frame. N is the size of the REF frame (each bit plane will have the size N).

$X_2 = \left(x_{2_1}, x_{2_2}, x_{2_N}\right)$, where $\left\{x_{2_i}\right\}$, $1 \le i \le N$ represent the "second" most significant bits obtained after quantification of the REF frame.

$Y_1 = \left(y_{1_1}, y_{1_2}, y_{1_N}\right)$, where $\left\{y_{1_i}\right\}$, $1 \le i \le N$ represent the most significant bits obtained after quantification of the frame $F1$. N is the size of the frame $F1$ (each bit plane will have the size N)

$Y_2 = \left(y_{2_1}, y_{2_2}, y_{2_N}\right)$, where $\left\{y_{2_i}\right\}$ $1 \le i \le N$ represent the "second" most significant bits obtained after quantification of the frame $F1$.

We define $X' = \left(x_{1_1}, x_{2_1}, x_{1_2}, x_{2_2}, x_{1_N}, x_{2_N}\right)$ and $Y' = \left(y_{1_1}, y_{2_1}, y_{1_2}, y_{2_2}, y_{1_N}, y_{2_N}\right)$. We calculate Z', where $Z' = X' \oplus Y'$ as the modulo-2 carried bit by bit. The binary vector Z' is representing the bits that have changed between the reference frame REF and the frame $F1$ i.e. the error between the two Most Significant bitplanes.

We perform a binary to decimal conversion of the Z' vector to obtain the Q-ary sequence ($Q = 2$) Z'' which will be encoded by the arithmetic encoder operating at symbol level. The encoded Q-ary sequence is then transmitted to the decoder.

The arithmetic decoder (operating at symbol level) decodes the received Q-ary sequence to obtain the vector $\tilde{Z}''$. A decimal to binary conversion is used to retrieve the vector $\tilde{Z}'$ and then we reconstruct $\tilde{Y}_1$ and $\tilde{Y}_2$ using $\tilde{Y}' = \tilde{X}' \oplus \tilde{Z}'$ where $\tilde{X}'$ is the vector obtained by intra decoding of the X' vector defined previously.

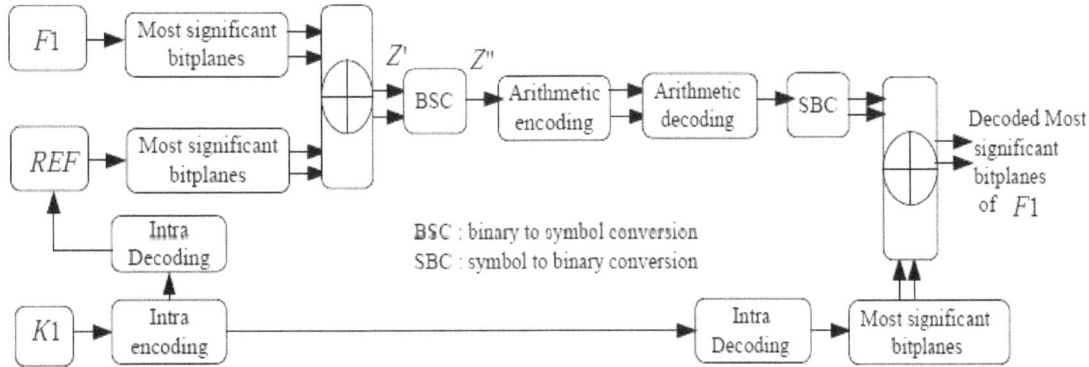

Figure 6: Most significant bitplanes encoding : Decimal level Arithmetic CoDec

• Exploitation of the spatial correlation

It is known that the video sequence contains a very large statistical redundancy, both in temporal and spatial domain. Furthermore the temporal correlation, we propose in this paragraph to exploit the spatial correlation within frames of the GOP.

- Spatial redundancy :

To exploit the spatial correlation, each frame F_k, $k = 1,2,3$ is divided into L blocks with size of 2×2 pixels each. Consider a 2×2 pixels block B : $B = P_1, P_2, P_3, P_4$. A decimal to binary conversion is applied to B, and we extract the two most significant bitplanes denoted respectively M and MS.

$$B_{binary} = (M_1, M_2, M_3, M_4; MS_1, MS_2, MS_3, MS_4).$$

- Temporal redundancy :

To exploit the temporal correlation, we generate the binary error between the frames REF and $F1$. Thus, the REF frame is divided into L blocks with size of 2×2 pixels each. A decimal to binary conversion is applied to each block. We consider $B1$ ($B1 = P_1', P_2', P_3', P_4'$) a block of the REF frame, the two most significant bitplanes of the block $B1$ are given by:

$$B1_{binary} = (M_1', M_2', M_3', M_4'; MS_1', MS_2', MS_3', MS_4').$$

We calculate the binary error $E = (E_1, E_2, ... E_8)$, where $E = B_{binary} \oplus B1_{binary}$ the sum modulo-2.

The binary vector $\{E\}$ which represents the bits that have changed between the binary vectors $B1_{binary}$ and B_{binary} is converted into a Q ary-symbol ($Q = 8$).

This process is applied to all the F_k's blocks. The sequence of symbols obtained by considering Q-ary the L blocks is encoded by an arithmetic encoder (operating at symbol level) to form the bitstream which is transmitted to the decoder.

The decoder decodes the bit stream to obtain $\tilde{E}$, and then rebuilt F_k. This process is described by Figure7.

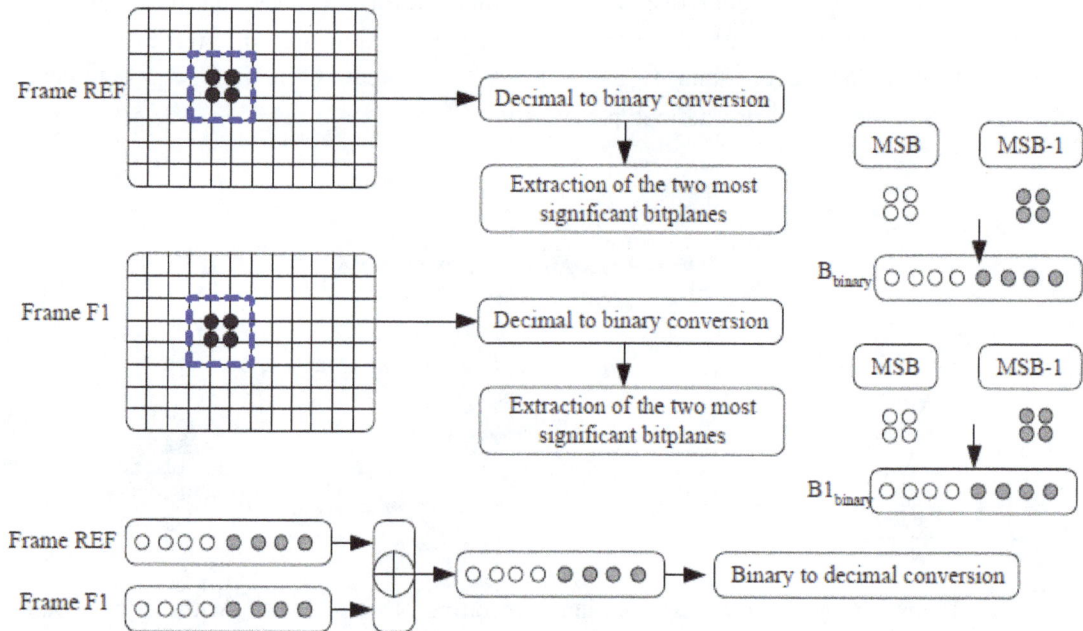

Figure 7: Most significant bitplanes encoding : Exploitation of the spatial correlation

4 Simulations Results

The system proposed in Section 3.1 was simulated in order to assess its performances. We consider the video sequence "Mother and Daughter" whose characteristics are shown in Table 1.

Evaluated frames	Spatial resolution	Temporal resolution
101	QCIF	30(fps)

Table   1:   characteristics of the video sequence "Mother and Daughter"

Each frame of the "mother and daughter" video sequence has a size of 144×176 pixels. To allow comparing the results obtained with our proposed work and those available in [5], the performance evaluation process is submitted to certain conditions:
- Only the luminance data is considered in the rate/distorsion performance evaluation.
- The key frames (K), represented in figure 2 by X_{2i-1} and X_{2i+1} are assumed to be losslessly available at the decoder.
- The turbo code used consists of two recursive systematic convolutional encoders of rate $1/2$, each one is represented by the generator matrix $[1 \frac{1 + D + D^2 + D^3 + D^4}{1 + D^4}]$. The parity bits are stored in the encoder buffer while the systematic part is discarded. The simulation set-up assumes ideal error detection at the decoder, i.e. the decoder can determine wether the bit-plane error rate, Pe, is greater than or less than 10^{-3}. If $Pe \geq 10^{-3}$ it requests for additional parity bits.

- A Laplacian distribution is used to model the residual between F_k, $k = 1,2,3$ and Y_{2i}. Each frame is characterized by a Laplacian parameter value α.

- The rate/distorsion plots only contain the rate and PSNR of the frames F_k, $k = 1,2,3$.
- The GOP size is considered equal to 3.
- For each sequence's frames F_k, $k = 1,2,3$, we gathered the quantized symbols to form an input block of length $N = 144 \times 176 = 25344$ quantified symbols. Over the resulting quantized symbol stream bitplane extraction is performed to generate 4 bitplanes noted (Y_1, Y_2, Y_3, Y_4) for the F_k, $k = 1,2,3$ frames (each bitplane has a length of N bits).

Least significant bitplanes processing:

For the least significant bitplanes Y_3 and Y_4, we consider the treatment presented in Figure 2. To achieve the rate compatibility for the turbo code, we devised an embedded puncturing scheme, with a puncturing pattern period of 22 parity bits. The parity bits are stored in a buffer, punctured, and transmitted upon decoder request via the feedback channel.

Most significant bitplanes processing:

For the most significant bitplanes Y_1 and Y_2, we follow the scheme shown in Figure 5. We generate, for each bitplane, the corresponding binary error by considering the bitplanes of the reference frame REF (as described in the previous section).

In a video sequence, the most significant bits tend to be constant if we consider two consecutive frames (except the case of plane change). Thus, the obtained binary errors vectors are highly correlated (long sequences of zero). We exploited this feature which highlighting the temporal redundancy, using an arithmetic coder. It will consider the correlation of binary error vector to generate a compressed data.

We simulated the distributed video coding scheme for a GOP size equal to 3 (as shown in Figure 3). This codec uses a turbo code to encode the different bitplanes generated after quantization. We represent, in the Table 2, rates needed for the transmission of the four bitplanes.

Subsequently, we simulated the codec shown in Figure 5 for the two most significant bitplanes. The bitplanes $MSB-2$ and $MSB-3$ (the two least significant bitplanes) still use the turbo encoder defined above. We opted for this scheme (which gathered together the arithmetic coder and the turbo coder) because we noticed that performances of the arithmetic coder decline for the last two bitplanes. This is explained by the fact that these bitplanes are low correlation compared to the previous ones (most significant bitplanes).

According to the results, we notice a gain which excess of 20% for the first bitplane. This gain decreases for the second bit plane to attempt 11.4%.

	MSB	MSB-1	MSB-2	MSB-3
Basic DVC (R1 kbps)	61.44	62.15	90.62	124.08
Arithmetic Codec (R2 kbps)	48.26	55.07	90.62	124.08
Δ Rate(%)= $\frac{R2-R1}{R1}$	-21.4	-11.4	0	0

Table 2: Results of the proposed scheme using an arithmetic encoder (binary level) for the most significant bitplanes

We focus on the two most significant bitplanes, and we simulate the scheme represented in Figure6. We perform a binary to decimal conversion of the error vector to obtain the Q-ary sequence ($Q=2$) which will be encoded by the arithmetic encoder operating at symbol level. The results achieved with this new scheme are shown in Table 4. We obtain a significant gain which reaches 38.85 kbps by comparing the pattern of reference presented in Figure 3. It represents 31% of the rate needed to decode the most significant bitplanes with the distributed video codec.

	MSB & MSB-1
Basic DVC(kbps)	123.59
Arithmetic Codec(kbps)	103.33
Arithmetic codec decimal (kbps)	84.71

Table 3: Results of the proposed scheme using an arithmetic encoder (symbol level) for the most significant bitplanes

We simulate the scheme represented in Figure7, the results achieved with this new scheme are shown in Table ??. We obtain a significant gain which reaches 51.71 kbps by comparing the pattern of reference presented in Figure 3. It represents 41.8% of the rate needed to decode the most significant bitplanes with a classic scheme (DCV codec).

	MSB & MSB-1
Basic DVC(kbps)	123.59
Arithmetic Codec(kbps)	103.33
Exploiting spatial and temporal correlation (kbps)	71.88

Table 4: Results of the proposed scheme using spatial and temporal correlation

5 Conclusion

In this paper, we have introduced a new scheme of distributed video coding, for a GOP size greater than 2, which exploit temporal correlation between frames to improve the performance of Wyner-Ziv decoding. We use an arithmetic encoder to exploit the correlation between the most significant bitplanes (MSB and $MSB-1$) of two successive frames. The obtained results exhibit a gain of 13kbps for the MSB bitplane and a gain of 7kbps for the $MSB-1$ bitplane.

Then, we improve the proposed scheme by using an arithmetic encoder which operates at symbol level. In fact, the two most significant bitplanes are grouped to form a sequence of symbols that will be encoded with the arithmetic encoder. Simulation results show a gain of 38,85 kbps in terms of rate compared with the basic solution presented in Figure3. This gain represents 31% of the total rate needed to transmit the MSB and the $MSB-1$.

The gain obtained by the scheme exploiting both spatial and temporal correlation attempt 51.71 kbps compared to the distributed video coding scheme.

As future work, it is planned to further enhance the RD performance of the codec by using others technique to more exploit the correlation in the video sequence.

References

[1] D.Slepian and J.K. Wolf,"Noiseless coding of correlated information sources", IEEE Trans.Inform.Theory, vol. IT-19,pp. 471-480, July 1973

[2] Wyner AD and Ziv J, "The rate distorsion function for source coding with side information at the decoder",IEEE Trans.Inform.Theory, vol. IT22,pp. 1-10, Jan. 1976

[3] Catarina BRITES, Advances on distributed video coding,PhD Thesis, The technical university of Lisbon, Dec 2005

[4] A.Aaron, S.Rane, E. Setton and B.Girod, "Transform-Domain Wyner-Ziv Codec for video", Visual Communications and Image Processing 2004. Edited by Panchanathan, Sethuraman; Vasudev, Bhaskaran. Proceedings of the SPIE, Volume 5308, pp. 520-528 (2004).

[5]   A.Aaron, R.Zhang, and B.Girod, "Wyner Ziv coding of motion video", in Proc. Asilomar Conference on Signals and Systems, Pacific Grove, California, Nov.2002

[6]   ISI/IEC International Standard 14496-10:2003, "Information Technology- Coding of audio-visual objects-Part10 : Advanced Video Coding"

[7] L.R.Bahl, J.Cocke, F.Jelinek, j.Raviv : "Optimal Decoding of Linear Codes for Minimising Symbol Error Rate", IEEE Trans.Inform.Theory, vol. 20,pp. 284-287, 1974

[8]   C.BERROU, A.GLAVIEUX : "Near Optimum Error Correcting Coding and Decoding : Turbo Codes", IEEE Transactions on communications, vol.44 NO.10, Oct. 1996

5

CONCATENATED CODING IN OFDM FOR WIMAX USING USRP N210 AND GNU RADIO

B. Siva Kumar Reddy [1] and B. Lakshmi [2]

Research Scholar[1], Associate Professor[2]
Department of Electronics and Communication Engineering
National Institute of Technology Warangal, Andhra Pradesh-506004, India.

ABSTRACT

A software Defined Radio (SDR) device employs a reconfigurable hardware (Universal Software Radio Peripheral-USRP) that may be programmed over-the-air or software (GNU Radio) to function under different Wireless standards. This paper analyzes the effect of various parameters such as channel noise, frequency offset, timing offset, timing beta, FLL (Frequency Lock Loop) bandwidth, Costas loop (phase) bandwidth, filter roll off factor and multiply const on OFDM signal in WiMAX physical layer with concatenated coding using SDR test bed. Concatenated coding is performed by suggesting RM coder and Convolutional coders as inner code and outer codes respectively. Moreover, bit error rate and symbol error rates performance are analyzed by varying bits per symbol, window size and modulation scheme. Results proved that BER and SER values are improved as modulation scheme size (M) is increased. OFDM signal transmission and reception is performed using USRP N210 and configured by GNU radio in the laboratory environment.

KEYWORDS

BER, GNU Radio, OFDM, SDR, USRP, WiMAX.

1. INTRODUCTION

Software Defined Radio (SDR) [1] is a term reinvented from software radio by Joseph Mitola in 1991, while recognizing the possibilities of re-configurability and re-programmability of radio systems. The idea behind software defined radio is to perform all signal processing functions with software instead of using dedicated circuitry. The most obvious benefit is the reduction in complexity and cost because of less hardware usage. An ideal SDR would have all the radio-frequency bands and modes determined software-wise, meaning it would comprise only of an antenna, DAC or ADC and a programmable processor. However, in practical systems, the RF front-end has to be enforced as well in order to support the receive/transmit mode. In this paper SDR is implemented by employing USRP (Universal Software Radio Peripheral) N210 [2] as hardware and GNU radio [3] as software platforms.

WiMAX (Worldwide Interoperability for Microwave Access) [4] is one of the most widely using broadband wireless access technologies based on the IEEE802.16 standard for Metropolitan Area Networks (MAN). WiMAX supports fixed and mobility services called as Fixed WiMAX (IEEE 802.16d) [4] and Mobile WiMAX (IEEE 802.16e-2005) [4] respectively. For mobile communications below 6 GHz frequencies have good propagation properties and are better suitable. 802.16 allows for several antennas to be employed at the transmitter and the receiver to provide a MIMO [5] system.

The field of channel coding [6] is pertained with transmitting a stream of data at as high a rate as possible over a given communications channel, and then decoding the original data reliably at the receiver, employing encoding and decoding algorithms that are executable to carry out in a given technology. The motivation for concatenating two coding schemes is to achieve large coding gains with affordable decoding complexity. In coding theory, concatenated codes [6] form a class of error-correcting codes that are gained by combining inner and outer codes. In this paper concatenated coding structured as Convolutional coding [6] as outer code and Reed Muller coding [7] as inner code.

The rest of the paper is structured as follows: Section 2 demonstrates the experimental setup of SDR with USRP N210 and PC (GNU Radio Companion (GRC)) and gives the clear explanation about USRP and GNU Radio platforms with specifications. Section 3 presents the WiMAX physical layer with working principles and explains each block which are used in GRC and for detailed information can refer ref [3]. Section 4 delivers observed experimental results and corresponding figures. Section 5 concludes the paper from the results obtained from Section 4.

2. EXPERIMENTAL SETUP

SDR comprises of RF section, IF section and baseband processing section. RF and IF sections are incorporated in USRP and baseband processing is performed in a PC using GNU radio companion (shown in Figure. 1). The USRP N210 [2] allows for high-bandwidth, high-dynamic range processing capability. This includes a Xilinx® Spartan® 3A-DSP 3400 FPGA (Field Programmable Gate Array), two 100 MS/s ADCs, two 400 MS/s DACs and Gigabyte Ethernet connectivity to flow information to and from host processors. The USRP N210 adds a larger FPGA than the USRP N200 [2] for additional logic, memory and DSP resources based demanding applications. All baseband signal processing (e.g. modulation, amplification, mixing, filtering etc.) is done in GNU Radio [3]. USRP can be reconfigured (in runtime also) to desired specifications in host computer by using GNU Radio. GNU Radio is a free software development toolkit that offers the signal processing runtime and readily available more than 100 processing blocks to implement software radios employing low-cost external RF hardware (USRP) and allows real time SDR applications [1]. In GNU Radio, signal processing blocks are written in Python and those are connected using C++ and both languages are communicated by SWIG (Simplified Wrapper and Interface Generator) interface compiler. Thus, the developer is allowed to accomplish real-time, high-throughput radio systems in a simple to-use, rapid-application development environment. In this paper all GNU Schematics (Signal flow graphs) are drawn for Mobile WiMAX specifications (FFT size=1024) [4].

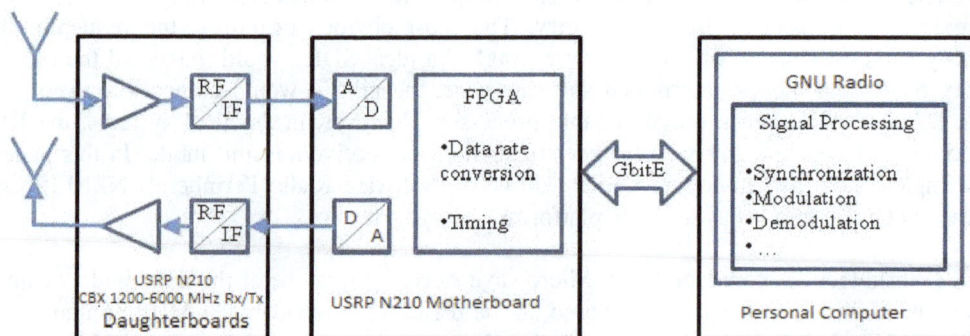

Figure 1: Software Defined Radio block diagram with USRP N210 and GNU Radio.

3. WIMAX PHYSICAL LAYER

The function of the PHY layer is to encode the binary digits that symbolize MAC frames into signals and to send and obtain these signals throughout the communication media. The WiMAX PHY layer is based on OFDM (Orthogonal Frequency Division Multiplexing)/OFDMA (Orthogonal Frequency Division Multiple Access) technologies [8] which are applied to enable high-speed data, video, and multimedia communications and is employed by a variety of commercial broad band systems. The WiMAX PHY layer (shown in Figure. 2) includes various functional stages: (i) Forward Error Correction (FEC): including; scrambling, concatenated encoding and interleaving (ii) OFDM modulation and (iii) Receiver synchronization.

The data flow processing through physical layer is described as follows. A signal with 6 GHz frequency is captured from the environment by using CBX daughterboard (in USRP N210) and GNU radio. The captured 6 GHz signal is passed to scrambler and it scrambles an input stream employing an LFSR (Linear Feedback Shift Register) [6]. This block influences on the LSB only of the input data stream, i.e., on an "unpacked binary" stream, and develops the same format on its output. The CCSDS encoder block [3] executes convolutional encoding [6] applying the CCSDS standard polynomial ("Voyager"). The input and output are an MSB first packed stream of bits and a stream of symbols *0* or *1* representing the encoded data respectively. Since the code rate is *1/2*, there will be *16* output symbols for every input byte. This block is planned for continuous data streaming, not packetized data. There is no provision to "*flush*" the encoder.

Data interleaving is used to increase efficiency of FEC by disseminating burst errors inserted by the transmission channel over a long time. The interleaving is determined by a two step permutation. First checks that adjacent coded bits are mapped onto nonadjacent subcarriers. The second permutation checks that adjacent coded bits are mapped alternately onto less or more significant bits of the constellation, thus eliminating long runs of lowly reliable bits. The first permutation is given by

$$m_k = (N_{cbps}/12)k_{mod(12)} + floor(k/12)$$
(1)

The second permutation is defined by [6],

$$j_k = s.floor(m_k/S) + [m_k + N_{cbps} - floor(12.m_k/N_{cbps})]_{mos(s)}$$
(2)

Where $k = 0, 1,..., N_{cbps}$; N_{cbps} is the number of coded bits per subcarrier, i.e., 1, 2, 4 or 6 for BPSK, QPSK, 16–QAM, or 64–QAM, respectively; k is the index of the coded bit before the first permutation; m_k is the index of that coded bit after the first and before the second permutation, and j_k is the index after the second permutation, just prior to modulation mapping. The receiver also does the reverse operation following the two step permutation using equations (3) and (4) respectively:

$$f_i = S.floor(j/S) + [j + floor(12.j/N_{cbps})]_{mod(s)}$$
(3)

$$S_j = (12.f_j - (N_{cbps}).floor(12.f_j/N_{cbps}));$$
$$j = 1, 2, ...N_{cbps}$$
(4)

In Reed-Muller Encoder [7] Only the first bit is used for in and output. m must be smaller than 31and r must be smaller than m. Reed–Muller codes are listed as *RM(d,r)*, where *d* is the order of the code, and r sets the length of code, n = 2^r. RM codes are related to binary functions on field $GF(2^r)$ over the elements {0,1}. RM(1,r) codes are parity check codes of length $n = 2^r$, rate $R = \dfrac{r+1}{n}$ and minimum distance $d_{min} = \dfrac{n}{2}$ [3].

OFDM block [3] generates OFDM symbols based on the parameters like *fft_length, occupied_tones, and cp_length* and *a type of modulation and etc* [8]. The transmitted signal voltage to antenna as a function of time during any OFDM symbol is defined as [3]

$$S(t) = \left(e^{j2\pi t f_c t} \sum_{k=-N_{used}/2}^{N_{used}/2} a_k . e^{j2\pi k \Delta f(t-T_g)} \right)$$

(5)

where t is time, elapsed since the beginning of the subject OFDM symbol with $0 < t < T_s$, a_k is a complex number ; the data to be transmitted on the carrier whose frequency offset index is k, during the subject OFDM symbol. It assigns a point in a QAM constellation, T_g is guard time, T_s is OFDM symbol duration including guard time, Δf is carrier frequency spacing. In the subsequence, carriers are distinguished by a carrier index; however in order to reconstruct the OFDMA signal, frequency offset index is required. OFDMA is a special case or multi user version of OFDM which offers frequency diversity by spreading out the carriers all over the applied spectrum. Frequency offset index is defined in terms of its carrier index by equation (6)

$$k_{foi} = \begin{cases} k_{ci} - N_{used}/2, & k_{ci} < N_{used}/2. \\ k_{ci} - N_{used}/2 + 1, & k_{ci} \geq N_{used}/2. \end{cases}$$

(6)

Where K_{foi} is carrier frequency offset index, K_{ci} is carrier index and N is number of used carriers. Chunks to symbols block [3] maps a stream of symbol indexes (unpacked bytes or shorts) to stream of float or complex constellation points. Input is stream of short and output is stream of float.

$$out[nD+k] = symbol\ table[in[n]D+k], k = 0, 1, ..., D-1.$$

(7)

The combination of *gr_packed_to_unpacked_XX* followed by *gr_chunks_to_symbols_XY* deals the general case of mapping from a stream of bytes or shorts into arbitrary float or complex symbols.

Poly phase resample block [3] accepts a single complex stream in and outputs a single complex stream out. As such, it needs no extra glue to deal the input/output streams. This block is supplied to be consistent with the interface to the other PFB (Poly phase filter banks) block [3]. PFBs are a very powerful set of filtering tools that can efficiently perform many multi-rate signal processing tasks. GNU Radio has a set of PFBs to be employed in all sorts of applications. This block consents a signal stream and performs arbitrary resampling. The resampling rate can be any real number r. The resampling is acted by constructing N filters where N is the interpolation rate. Then D can be defined as, $D = floor(N/r)$. Using N and D, rational resampling is performed. where N/D is a rational number close to the input rate r where $i+1 = (i + D)$ % N. To acquire the arbitrary rate, interpolation between two points is required. For each value out, an output from the current filter, i, and the next filter $i+1$ are considered and then linearly interpolate between the two based on the real resampling rate.

Table 1. SNR estimation with variation in multiply const value.

Multiply Constant value	Estimated SNR Value
1	1.00234
10	10.6085
50	52.809883
100	108.647

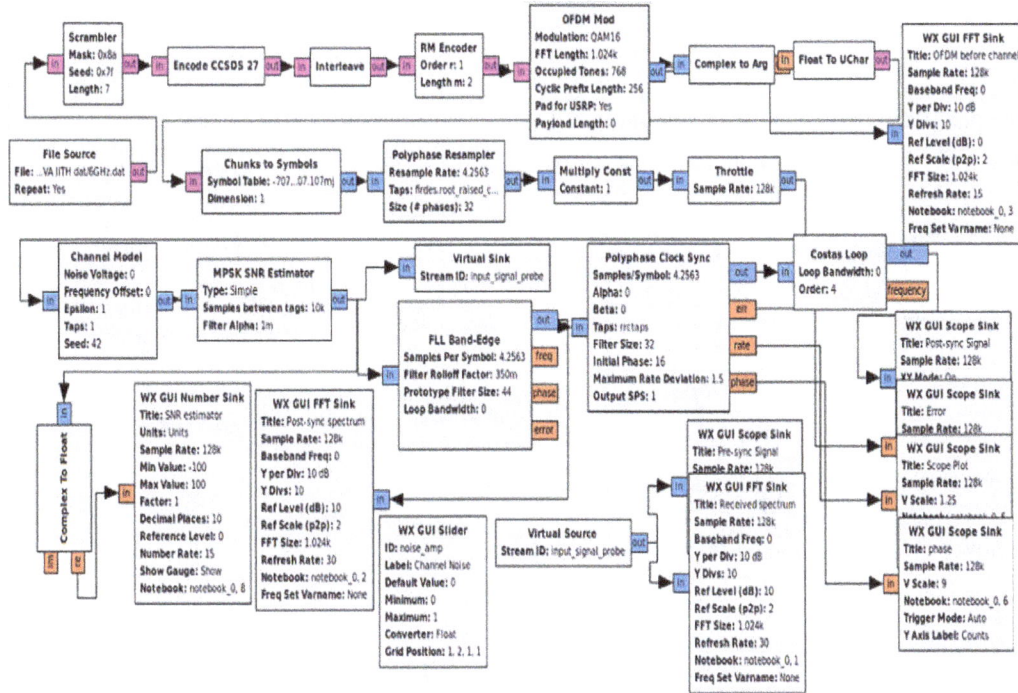

Figure 2: GNU schematic for OFDM signal transmission and reception over virtual source and sink.

Channel Model block [3] implements a basic channel model simulator that can be applied to help evaluate, design, and test various signals, waveforms, and algorithms. This model appropriates the user to set the voltage of an AWGN noise source, a (normalized) frequency offset, a sample timing offset, and a noise seed to randomize the AWGN noise source [9]. Multipath can be estimated in this model by using a FIR filter representation of a multipath delay profile. MPSK SNR estimator [3] is block for computing SNR of a signal. This block can be employed to monitor and retrieve estimations of the signal SNR. It is designed to work in a flow graph and passes all incoming data along to its output. Estimated SNR value is increased as multiply const block value increased (shown in Table 1) i.e. multiply const block acts as an amplifier in the schematic (See Fig. 2) .

The frequency lock loop [3] derives a band-edge filter that covers the upper and lower bandwidths of a digitally modulated signal. The bandwidth range is determined by the excess bandwidth (e.g., roll off factor) [3] of the modulated signal. The placement in frequency of the band-edges is determined by the oversampling ratio (number of samples per symbol) and the excess bandwidth. The size of the filters should be fairly large so as to average over a number of symbols. The FIR filters are employed here because the filters have to have a flat phase response over the entire frequency range to allow their comparisons to be valid. It is very important that the band edge filters be the derivatives of the pulse shaping filter, and that they be linear phase. Otherwise, the variance of the error will be very large. Poly phase clock sync block performs timing synchronization for PAM signals by minimizing the derivative of the filtered signal, which in by turn maximizes the SNR and minimizes ISI [8].

The Costas loop [10] can have two output streams: stream 1 is the baseband I and Q; stream 2 is the normalized frequency of the loop. Digital Costas loop consists of Direct Digital Synthesizer (DDS), Low Pass Filter (LPF) and a Phase Discriminator (PD) and a Loop Filter (LF). Suppose that the input signal is a baseband signal modulated by the intermediate frequency carrier signal is given by [10]

$$x(t) = M(t).cos(w_c t) \qquad (8)$$

The in phase and quadrature branch outputs of local DDS s are as follows respectively

$$V_{oi} = cos(w_c t + \Delta\phi) \qquad (9)$$

$$V_{oi} = sin(w_c t + \Delta\phi) \qquad (10)$$

Where Δ is the phase difference between input signal and local signal of DDS. Then the multiplier outputs of in phase and quadrature branch are as follows

$$Z_i(t) = M(t).cos w_c t.cos(w_{ct} + \Delta\phi) \qquad (11)$$

$$Z_q(t) = M(t).cos w_c t.sin(w_{ct} + \Delta\phi) \qquad (12)$$

After low pass filtering, the corresponding outputs are

$$y_i(t) = \frac{1}{2}k_{l1}M(t)cos(\Delta\phi) \qquad (13)$$

$$y_q(t) = \frac{1}{2}k_{l2}M(t)sin(\Delta\phi) \qquad (14)$$

Where k_{l1}, k_{l2} are low pass filter coefficients. After $y_i(t)$ and $y_q(t)$ passed through phase discrimination and loop filter, following equation is obtained

$$V_c(t) = \frac{1}{8}k_p k_{l1} k_{l2} sin(2\Delta\phi) = k_d sin(2\Delta\phi) \qquad (15)$$

4. CHANNEL CODING

In digital communications, a channel code is the term relating to the forward error correction code and interleaving in communication and storage where the communication media is recognized as a channel. The channel code is utilized to defend data sent over it for storage or recovery even in the orientation of noise (errors). Channel coding [6] is referred to process in both transmitter and receiver of a digital communications framework. Channel coding is made out of three techniques, for example Randomization, FEC (Forward Error Correction) and Interleaving.

5. EXPERIMENTAL RESULTS

A real time Software Defined Radio (SDR) is developed (as shown in Figure 3) by using a laptop with 8 Giga Bytes of RAM and an Intel ® Core™ i5-3210M CPU clocked at 2.50 GHz. The integrated 1000Base-T Ethernet interface was connected to the USRPN210, equipped with the CBX daughterboard which is a full-duplex, wide band transceiver that extends a frequency band from 1.2 GHz to 6 GHz with a instantaneous bandwidth of 40 MHz (set up shown in Figure. 4). The CBX can serve a wide variety of application areas, including Wi-Fi research, cellular base stations, cognitive radio research, and RADAR. Required OFDM parameters for WiMAX specifications are mentioned in Table 1. Figure. 4 presents the encoding-decoding block diagram of the concatenated coding system for BER analysis [11] over air using USRP source and sink. RM coder (Reed-Muller code) [7] and CCSDS encoder (Convolutional coder) [3] are employed as inner code and outer code respectively. The concatenated OFDM signal is transmitted by USRP N210 RF front end by using TX/RX antenna and received by RX2 antenna over air (see Figure 3) in the lab environment. BER performance is analyzed by varying bits per symbol and window size in Error rate block and the results are observed in Table 3. It can be concluded that BER performance is improved as number of bits per symbol is increased and varies with window size. As modulation scheme size increases BER also increased (Observe Table 5) which is not desirable. Hence, while preferring a type of modulation scheme, various parameters have to be taken in to consideration.

In an OFDM transmission, we know that the transmission of cyclic prefix does not carry 'extra' information in Additive White Gaussian Noise channel. The signal energy is spread over time $T_d + T_{cp}$ whereas the bit energy is spread over the time T_d i.e.

$$E_s(T_d + T_{cp}) = E_b.T_d$$
$$E_s = \frac{T_d}{T_d + T_{cp}}.E_b$$

$$(16)$$

The relation between symbol energy and the bit energy is given by

$$\frac{E_s}{N_0} = \frac{E_b}{N_0}.\left(\frac{nDSC}{nFFT}\right)\left(\frac{T_d}{T_d + T_{cp}}\right)$$

$$(17)$$

Expressing in decibels

$$\frac{E_s}{N_0}dB = \frac{E_b}{N_0}dB + 10log_{10}\left(\frac{nDSC}{nFFT}\right) + 10log_{10}\left(\frac{T_d}{T_d + T_{cp}}\right)$$

$$(18)$$

Figure 3: Software Defined Radio development test bed.

Table 2. Experimental parameters defined

Parameters	Values
FFT size (NFFT)	1024
Occupied Tones	768
Sampling rate	10.66667M
Center Frequency	2.48 GHz
Convolutional Code	1/2
Cyclic Prefix length	256
Useful symbol duration	91.43 μs
Carrier spacing (1/Tu)	10.94 KHz
Guard time ($T_g = (1/4) * T_u$)	11.43 μs
OFDM symbol duration	102.86 μs
Mapping Schemes	BPSK, QPSK, 16QAM, 64QAM and 256QAM

Table 3. BER analysis with Bits per symbol and window size.

Bits per symbol	Window size	BER
1	10	3.5000000
1	1000	3.3710000525
1	10^6	3.4379618168
4	10	0.8594650625
4	1000	0.84799997
4	10^6	0.8594650625
8	10	0.4250000119
8	1000	0.42687499
8	10^6	0.4296149015

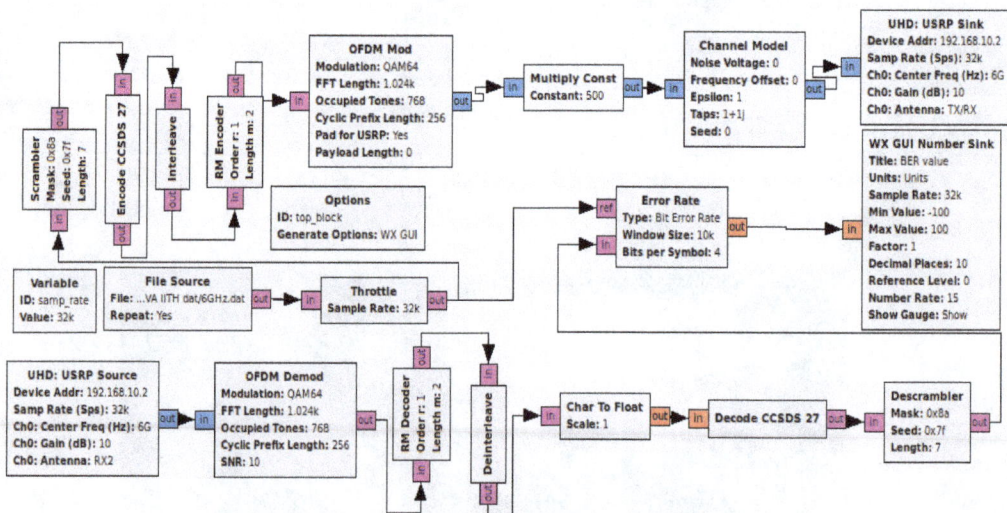

Figure 4: GNU Schematic for OFDM transmission and reception over USRP source and sink for BER analysis.

Table 4. BER and SNR analysis with Modulation scheme.

Modulation scheme	BER	SER
BPSK	0.8487750292	0.9564999938
QPSK	0.8616499901	0.9578999877
8PSK	0.8636000156	0.9592999816
16QAM	0.8638749719	0.9585999846
64QAM	0.8650249839	0.9577000141
256QAM	0.8660741590	0.9564999938

In order to evaluate the error probability, without loss of generality, paper focuses on the signal received on the first subcarrier, dropping the block index 1 for the sake of simplicity. A scaled version of the decision variable is given by

$$Z_1 = \lambda_1 Z_{EQ,1} = m_1 \lambda_1 S_1 + \sum_{n=2}^{N} m_n \lambda_n S_n + v_1$$

(19)

Where $Z_{EQ,1} = Z_{EQ}[l]_1$, $S_n = S[l]_n$, $v_1 = \square[l]^{-1} v[l]_1$ and

$$m_n = [M]_{1,n} = \frac{sin(\pi(n-1+\epsilon))}{N sin \dfrac{\pi(n-1+\epsilon)}{N}} . e^{j\pi \frac{N-1}{N}(n-1)}$$

(20)

represents the ICI coefficient due to the nth subcarrier for $n=2,,,,N$, and the attenuation factor of the useful data when n=1.

A possible approach to obtain BER (or equality, SER) consists of two steps. Firstly calculate the conditional bit error probability $P_{BE}(S,\lambda)$ that depends on the symbols in $S=[s_1,,,,,s_N]^T$ and on the channel amplitudes in $\lambda=[\lambda_1,....\lambda_N]^T$. Successfully, $P_{BE}(S,\lambda)$ should be averaged over the joint probability density function (pdf) $f_{S,\Lambda}(S, \lambda)= f_S(S) f_\Lambda(\lambda)$ of the symbols and channel amplitudes is given by

$$BER = \int_{S,\lambda} f_S(S) f_\Lambda(\lambda) dS d\lambda$$

(21)

Figure. 2 shows the GRC schematic drawn for analysis of channel noise effect on concatenated OFDM signal over virtual sources and sink. Figures. 5, 6 & 7 present the OFDM signal, post synchronized spectrum and post synchronized signal before applying channel noise respectively. Figure 7 shows the various parameters such as channel noise, frequency offset, timing offset, timing beta, FLL (Frequency Lock Loop) bandwidth, Costas loop (phase) bandwidth and filter roll off factor which are effecting the transmitted signal. When one of these parameters is varied, received/synchronized signals are changed accordingly. Out of band radiation has become indistinguishable to the in band radiation due to channel noise (shown in Figure 8). Received and synchronized signals are disturbed by the channel noise and became glazed over (shown in Figures 9 & 10). Received resembled signal after Costas loop is shown in Figure 11. Post synchronized signal is effected by timing alpha and Costas loop bandwidth is shown in Figures 12 & 13. The phase variation in the signal is shown in Figure 14 with the effect of timing beta.

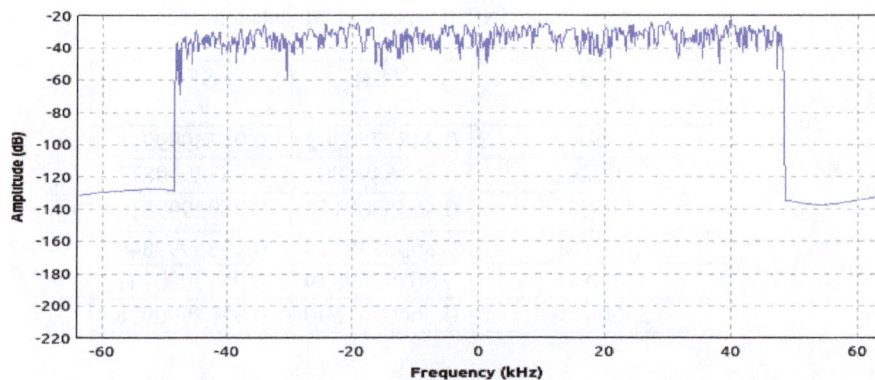

Figure 5: OFDM signal before passes through channel.

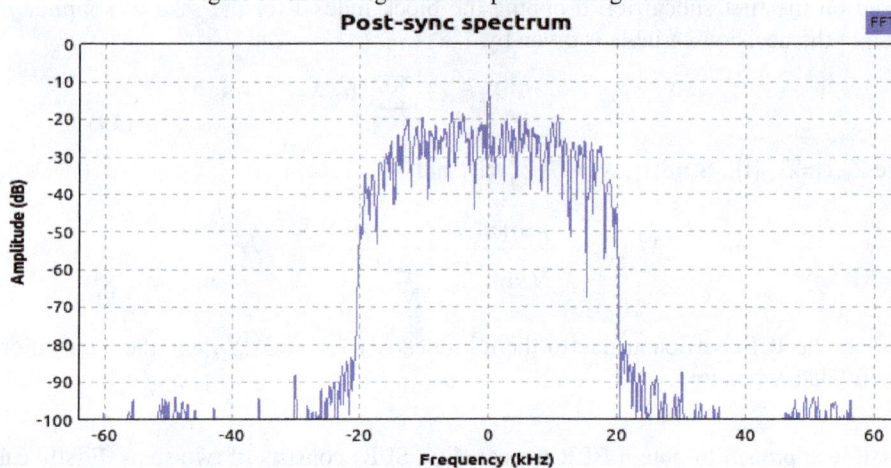

Figure 6: Post synchronized spectrum without channel effect.

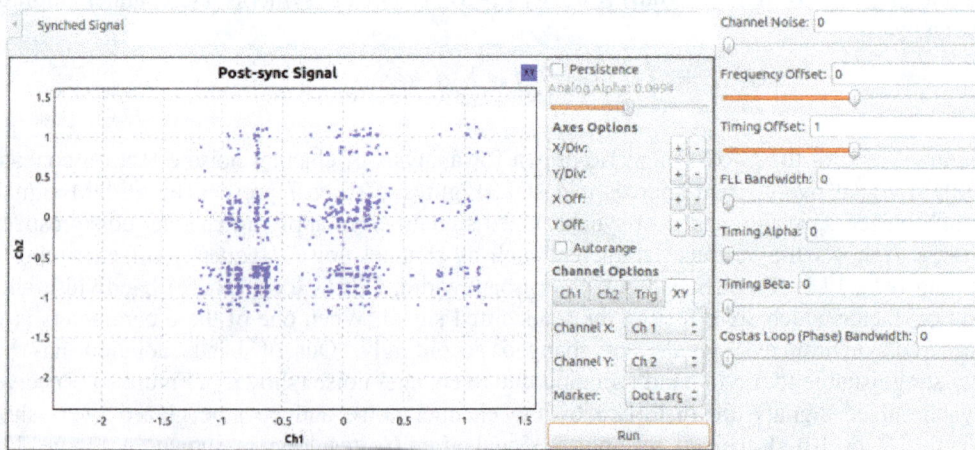

Figure 7: Post synchronized spectrum without channel effect.

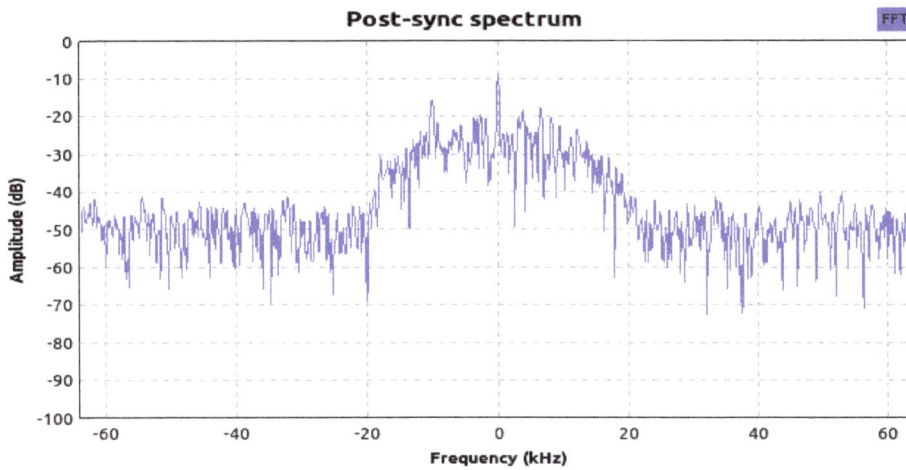

Figure 8: Post synchronized spectrum with effect of channel noise is 100m units.

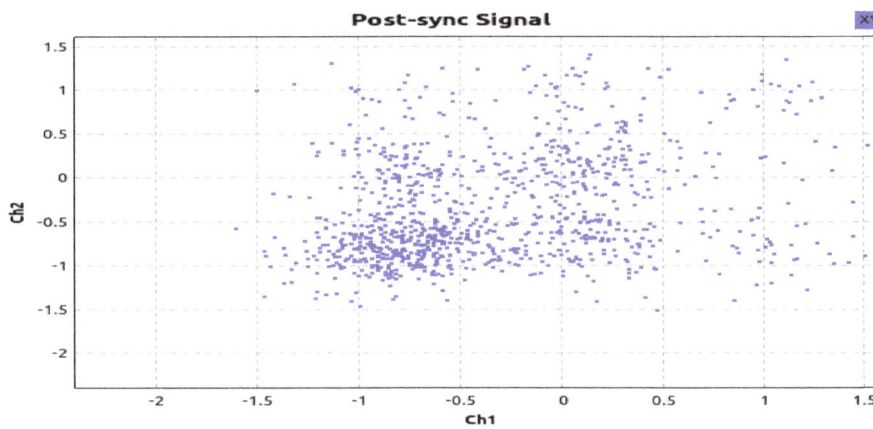

Figure 9: Post synchronized signal with effect of channel noise is 400m units.

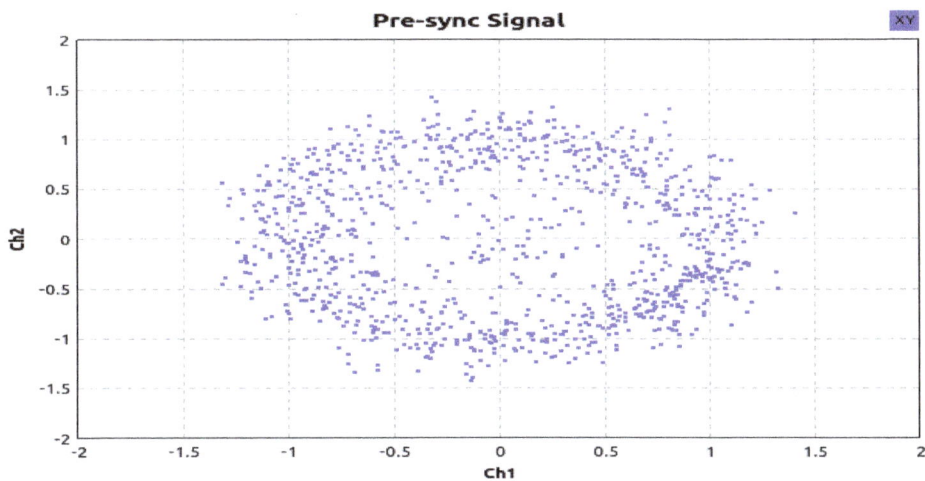

Figure 10: Received signal with effect of frequency offset is 11m units.

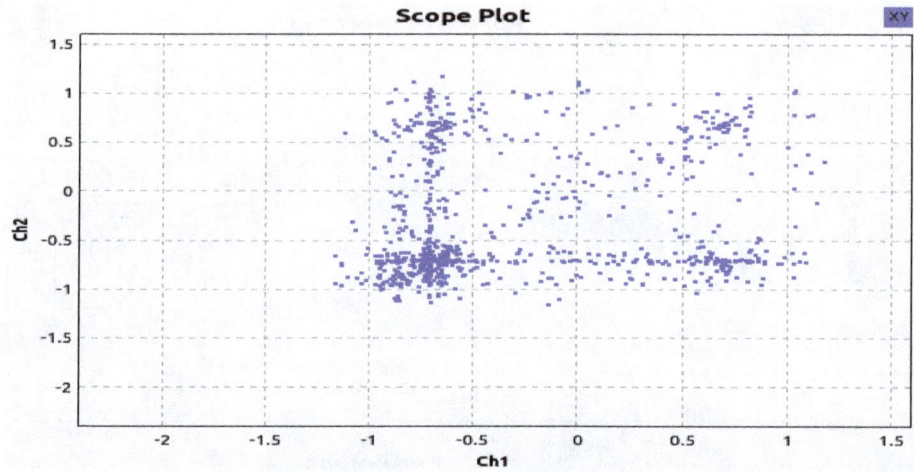

Figure 11: Received signal with effect of frequency offset is 11m units.

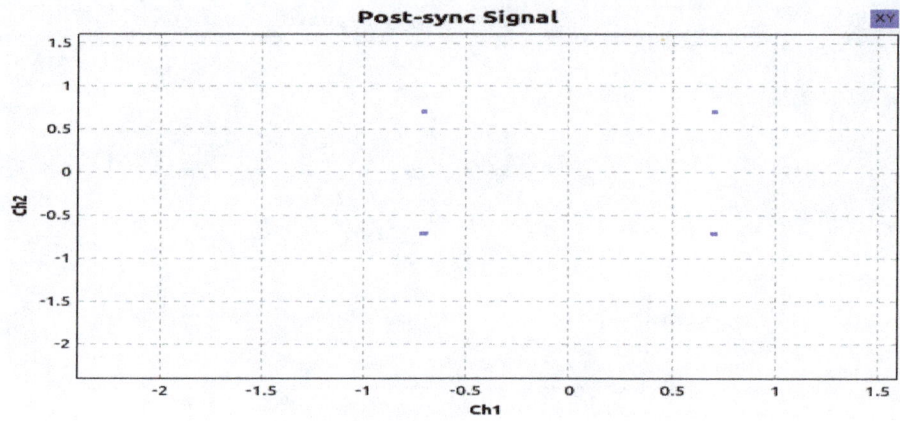

Figure 12: Post synchronized signal with effect of timing alpha is 20m units.

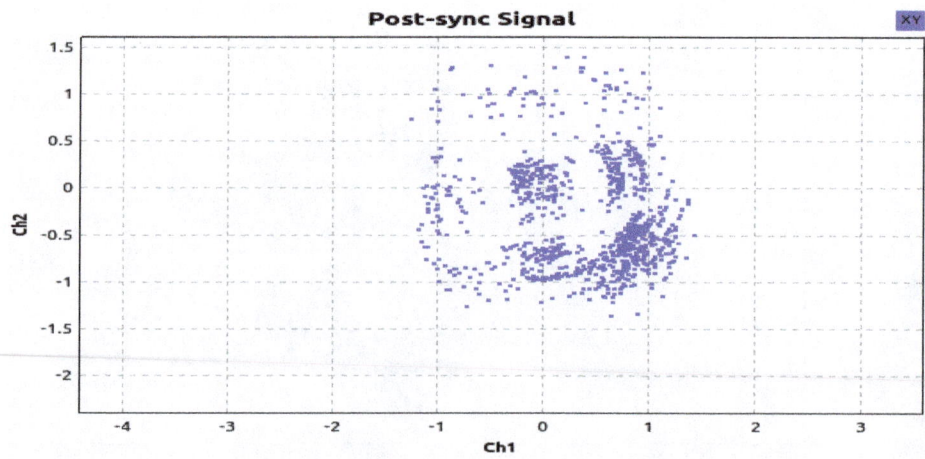

Figure 13: Post synchronized signal with effect of Costas loop (phase) bandwidth is 10m units.

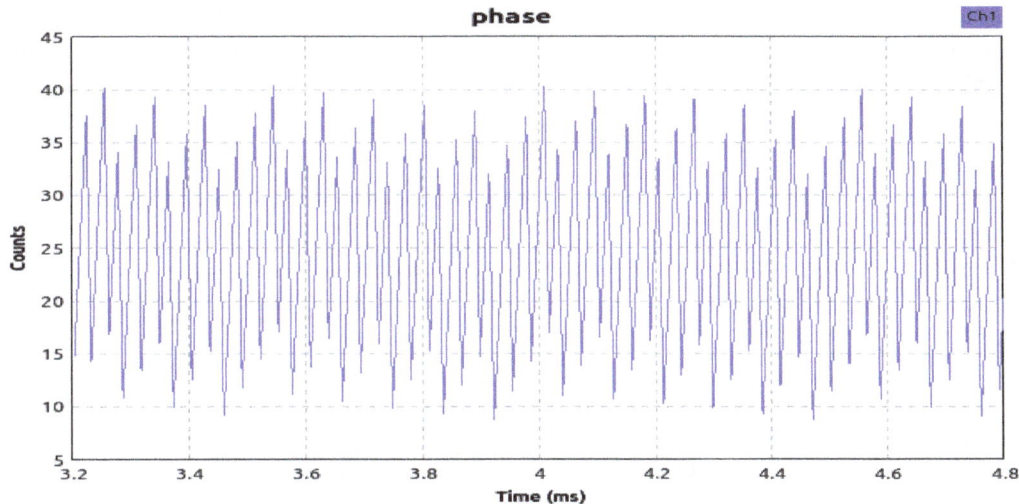

Figure 14: Phase variation with effect of timing beta is 10m units.

6. CONCLUSION

This paper shows the advancement of software defined radio (SDR) utilizing USRP N210 fittings and GNU Radio programming. OFDM/OFDMA based physical layer is executed with concatenated coding by considering RM code as internal code and Ccsds code as external code for Mobile WiMAX determinations at different (i) modulation schemes (ii) Channel noise levels (iii) frequency offsets (iv) costas loop bandwidth and (v) phase variation. As a result of the comparative study, it was found that: when channel conditions are poor, energy efficient schemes such as BPSK or QPSK were used and as the channel quality improves, 16-QAM or 64-QAM was used. It adjusts the modulation method almost instantaneously for optimum data transfer, thus making a most efficient use of the bandwidth and increasing the overall system capacity. Out of band radiation has gotten unclear to the in band radiation because of channel noise. Experiments validate the effectiveness of the proposed scheme in real time.

In this work, the measurement setup was somewhat idealized, since all measurements were conducted in a shielded environment. Future work should also include more realistic scenarios, such as interference from other secondary users or neighbouring frequency bands. More practical and better use of varying gain control should also be considered.

REFERENCES

[1] Luiz Garcia Reis, A.; Barros, A.F.; Gusso Lenzi, K.; Pedroso Meloni, L.G.; Barbin, S.E., "Introduction to the Software-defined Radio Approach," Latin America Transactions, IEEE (Revista IEEE America Latina) , vol.10, no.1, pp.1156,1161, Jan. 2012.

[2] Ettus, Matt. "USRP User's and Developer's Guide," Ettus Research LLC, 2005.

[3] Radio, G. N. U. "The gnu software radio." Available from World Wide Web: https://gnuradio. Org, 2007.

[4] Andrews, Jeffrey G., Arunabha Ghosh, and Rias Muhamed, Fundamentals of WiMAX: understanding broadband wireless networking. Pearson Education, 2007.

[5] Reddy, B.S.K.; Lakshmi, B., "Channel estimation and equalization in OFDM receiver for WiMAX with Rayleigh distribution," *Advanced Electronic Systems (ICAES), 2013 International Conference on,* pp.337,339, 21-23 Sept. 2013

[6] Pin-Han Chen; Jian-Jia Weng; Chung-Hsuan Wang; Po-Ning Chen, "BCH Code Selection and Iterative Decoding for BCH and LDPC Concatenated Coding System," Communications Letters, IEEE , vol.17, no.5, pp.980,983, May 2013.

[7] Reddy, Kumar, B. Siva, and B. Lakshmi. "Channel Coding and Clipping in OFDM for WiMAX using SDR." *International Journal on Recent Trends in Engineering & Technology*, 2013.

[8] Reddy, B. Siva Kumar, and B. Lakshmi. "Adaptive Modulation and Coding in COFDM for WiMAX Using LMS Channel Estimator." , 2013.

[9] Reddy, B.S.; Boppana, L.; Srivastava, A.; Kodali, R.K., "Modulation switching in OFDM for WiMAX through Rayleigh fading channel using GNU radio," *Advanced Electronic Systems (ICAES), 2013 International Conference on* , pp.331,333, 21-23 Sept. 2013

[10] Guo, Si Yu, Dong Hai Qiao, and He Ming Zhao. "Implementation of Costas Loop for BPSK Receivers Using FPGA." Applied Mechanics and Materials 263, pp.990-993, 2013.

[11] Rugini, Luca, and Paolo Banelli. "BER of OFDM systems impaired by carrier frequency offset in multipath fading channels." Wireless Communications, IEEE Transactions on 4.5: pp.2279-2288, 2005.

6

State-Space Approaches for Modelling Reduction of Pilot Symbol Assisted Modulation and Their Impact on the Channel Estimation

Hayder J. Albattat[1], Haider M. AlSabbagh[2], and S. A. Alseyab

Department of Electrical Engineering, Basarah University, Basrah, Iraq
[1]`saidhaider75@yahoo.com`
[2]`haidermaw@ieee.org`

ABSTRACT

This paper outlines the use of a balance truncation algorithm to design low order infinite impulse response (IIR) interpolator for pilot symbol assisted modulation. A state space model is developed to optimize filter coefficients. The proposed design for the filter has a frequency response is quite stable. This approach is highly beneficial for systems employing fading channel estimation. A comparison is given for the achieved results to that with ideal case.

KEYWORDS

FIR, IIR, Fading channels, PSAM

1. INTRODUCTION

In a world of dramatic changing technology, people need more and more requirements to communicate and get connected with each other and facilitate their life. These approaches may relate with appropriate and timely access to information regardless of the location of people or type of information like scientific collaboration, telemedicine, and real-time environment monitoring. These applications require access to high-bandwidth real time data, images, and video captured from remote sensors such as satellite, radars, and echocardiography [1]. The task at the receiving end of a communication system is to decode the received signal and produce a bit stream that matches the original transmission ones. However, due to the distortion caused by the channel, this process cannot be fulfilled directly. Therefore, a suitable way should be finding to overcome such issue to adjust the received signals before starting the demodulation process. One of the efficient methods is Pilot Symbol Assisted Modulation (PSAM). This method is used to reduce the effects of the fading in mobile communication systems [2-4]. The basic concept of the PSAM is to multiplex training symbols known to the receiver into the data stream. These pilot symbols and the specific multiplexing scheme are known at the receiver and can be exploited for channel estimation, receiver adaptation, and optimal decoding. The receiver uses the pilot symbol to estimate the channel state information (CSI) [5]. In recent years, reduction techniques for linear dynamical systems have drawn a considerable attention. Several interpolation methods have been proposed for PSAM, including low-pass sinc interpolation [2], and optimal Wiener interpolator [6]. Also, there are numerous models have been proposed to analyse the performance in the frequency domain [7]. Moore, in [8], developed a state-space concept based on measurements of controllability and observability in certain directions of the state space coordinates. It is known that grammians are not invariant under coordinate transformation. On the other hand, Beliczynski et. al., [9] presented a no minimal realization method for obtaining reduced order model for multivariable system. They have shown that such method encompasses the methods of aggregation (eigenvalue preservation) and moments matching. This paper presents a proposal bases on representing the goal function as a sum of orthonormalized complex decaying exponential. In relative large order the state space concepts in reduction implementation is often simplified, for its small matrix computation requirements. Also, a method to reduce the complexity of the wireless estimator is introduced. This approach is based on an efficient reduction form high order estimator to a low order one with an acceptable tolerance. The paper is organized as follows: Data format is illustrated in section 2 and modelling with Balance truncation is given in section 3 then the obtained results are presented in section 4, finally the conclusion is drawn to section 5.

2. Data format:

The data is formatted into frames of symbols. The first symbol in each frame occupied for the pilot as shown in Fig.1, where the NP is the total number of symbols per frame.

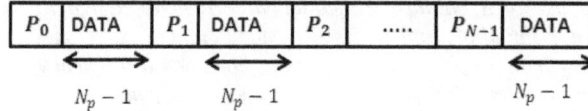

Fig. 1 Frame format

The pilot vectors P_i for i=0, 1,...., N-1 represent the input to the interpolator. The receiver estimates the fading at the i-th data symbol time in the n-th frame from the nearest NP pilot symbols, i.e., the receiver uses $\lfloor (N_P - 1)/2 \rfloor$ pilot symbols from previous frames, the pilot symbol from the current frame, and the pilot symbols $\lfloor N_P/2 \rfloor$ from the subsequent frames is illustrated in Fig.2. Thus, the estimated fading is given by:

$$\tilde{z}_n^{\ i} = \sum_{k=-\lfloor (N_P-1)/2 \rfloor}^{\lfloor N_P/2 \rfloor} f_k^{\ i} P_k \qquad (1)$$

where $f_k^{\ i}$ is the coefficient of the interpolator.

Sinc interpolation is used in this work due to its simplicity with the PSAM. The interpolation coefficients are computed from the sinc function as:

$$f_k^{\ i} = sinc(\frac{i}{N_p} - k) \qquad (2)$$

where $k = -\lfloor (N_P - 1)/2 \rfloor, ..., \lfloor N_P/2 \rfloor$. The sinc interpolator acts finite impulse response filter (FIR). FIR filters are simple to design and are guaranteed to be bounded input-bounded output (BIBO) stable. By designing the filter taps to be symmetrical about the centre tap position, the FIR filters have linear phase characteristics. The principle that is required throughout the pass-band of the filter response to preserve the shape of a given signal within the pass-band. Such property is desired for many applications such as in music and video processing [10].

Another type of the digital filter is (IIR filter) which occupies much lower implementation complexity than the FIR filter since it appears with a smaller order[10]. IIR filters suffer from a nonlinear-phase response, which limits their applications in 1-D system. To compensate this phase nonlinearly some additional phase equalizations circuit are required. This intern will add some extra complexity to the system implementation.

Usually the problem of approximation of a given system by lowering the order (requiring fewer memory elements system) has been surfaced in signal processing and control literature [9,11,12]. So, the yielding system after approximation support suppressing the complexity with keeping the same magnitude and phase responses as those of high order system (full order system) which are often desirable in practice. The linear model order reduction problem is given as "a Linear time invariant" (LTI) system G(z) of order N > n is given. It is desired to find an equivalent LTI system $\tilde{G}(z)$ of order no greater than n such that $\left\| G(z) - \tilde{G}(z) \right\|$ is as small as possible, where $\left\| . \right\|$ denotes some distance given by [8] and [2].

For

$$G(z) - \tilde{G}(z) = e(z) = \sum_{i=0}^{n} \varepsilon_i z^{-i} \tag{3}$$

where

$$\varepsilon_i = \frac{1}{2\pi j} \oint_c e(z).z^{i-1} dz \tag{4}$$

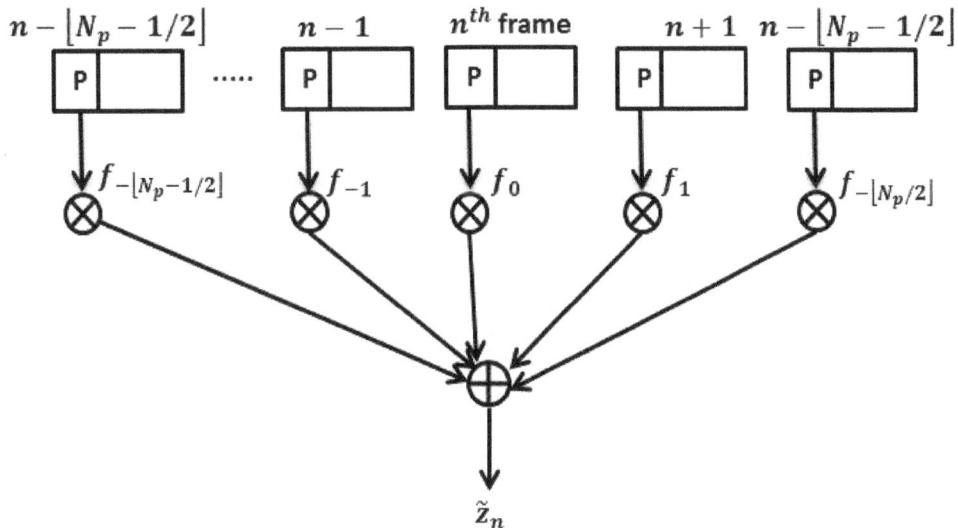

Fig. 2 Fading calculation in PSAM.

3. Balanced realization and reduction model:

3.1 Balanced realization:

The method of balanced realization is widely used in the context of realization theory for model of reduction construction of linear systems [11, 13]. This approach requires only standard matrix computations. The state-space description of a stable system is transformed to balanced coordinate via nonsingular matrix tool (T) satisfies that the observability and controllability grammians are equal and diagonal (balanced) [11, 12, 13]. Let $(A, B, C, D)_N$ represents a full-order 1-D dynamic stable system described by[12]:

$$x(k+1) = \mathbf{A} \, x(k) + \mathbf{B} \, u(k)$$
$$y(k) = \mathbf{C} \, x(k) + \mathbf{D} \, u(k) \tag{5}$$

where N is the order of realization with z-domain transfer function F(z) that is:

$$F(z) = C(ZI - A)^{-1}B + D \tag{6}$$

The controllability and observability grammians are then defined, receptively, as [15]:

$$W_c = \sum_{k=0}^{\infty} A^T B B^T (A^T)^k \tag{7}$$

and

$$W_o = \sum_{k=0}^{\infty} A^T C C^T A^T \tag{8}$$

The grammians are the unique positive defined solution for the Lyapunov equations as [14]:

$$AW_c + W_c A^T = -BB^T \tag{9}$$

then,

$$A^T W_o + W_o A = -C^T C \tag{10}$$

Under nonsingular state transfer matrix via T.

$$F(z) = \overline{C}(ZI - \overline{A})^{-1}\overline{B} + \overline{D} \tag{11}$$

where,

$$\overline{A} = T^{-1}AT, \overline{B} = T^{-1}B, \quad \overline{C} = CT \text{ and } \overline{D} = D \tag{12}$$

3.2 Hankel operator:

Over the past few decades, the Hankel operator has received a great deal of attention due to its vital role in control and filter model reduction [15]. Consider the proper transfer function:

$$G(z) = \frac{N(z)}{D(z)} = \frac{\beta_o + \beta_1 z^{-1} + \dots + \beta_n z^{-n}}{\alpha_o + \alpha_1 z^{-1} + \dots + \alpha_n z^{-n}} \tag{13}$$

where β_i's and α_i's (i=0,1,2,…,n) are constants. It can be expand into an infinite power series of descending power of z^{-1} as:

$$G(z) = c_0 + c_1 z^{-1} + c_2 z^{-2} + \dots \tag{14}$$

where the coefficients $\{c_i, i = 0,1,2,...\}$ are called Markov parameters. These parameters can be obtained recursively form A, B, C, and D of its state-space representation as:

$$c_i = CA^{i-1} \quad for \quad i = 1,2,3,... \tag{15}$$

and

$$D = c_0 \tag{16}$$

We define an infinite block-Hankel matrix, and denoted by $\Gamma\{G(z)\}$, as

$$\Gamma\{G(z)\} = \begin{bmatrix} c_1 & c_2 & c_3 & \ldots \\ c_2 & c_3 & \ldots & \\ c_3 & c_4 & \ldots & \\ & & \cdot & \\ & & \cdot & \end{bmatrix} \tag{17}$$

$\Gamma\{G(z)\}$ is formed from coefficients, representing the impulse response. According to Kronecker's theorem [8], the proper form of the transfer function G(z) has degree n if and only if:

$$R[\Gamma G(z)\}] = \deg[\,G(z)] \tag{18}$$

The Hankel matrix of Eqn. (17) can be written as:

$$\Gamma\{G(z)\} = QP \tag{19}$$

where P and Q are the controllability and observability matrices, respectively. And defined in state-space coordinates (A, B, C) as [8]:

$$P = [B: AB : \ldots : A^k B : \ldots] \tag{20}$$

$$Q = [C^T : A^T C^T : \ldots : (A^T)^k C^T : \ldots]^T \tag{21}$$

Since the solution of Eqn. (9) can be written as:

$$W_c = BB^T + ABB^T A^T + \ldots + A^k BB^T (A^T)^k + . \tag{22}$$

then

$$W_c = QP^T \tag{23}$$

and the similar one can show that:

$$W_o = Q^T P \tag{24}$$

The Hankel singular values of $G(z)$ are the singular values of $\Gamma\{G(z)\}$ and following holds true [15]:

[The ith singular value of $\Gamma\{G(z)\}]^2$ = The ith singular value of the product

$$W_c W_o = [\text{The } i^{th} \text{ singular value of } G(z)]^2 \tag{25}$$

The filter $G(z)$ may be represented as a set of differences equations as given in (5)

where,

$$A = \begin{bmatrix} 0 & 0 & \cdots & 0 & 0 \\ 1 & 0 & \cdots & 0 & 0 \\ & & \vdots & & \\ 0 & 0 & \cdots & 1 & 0 \end{bmatrix} ,$$

$$B = \begin{bmatrix} 1 \\ \vdots \\ 0 \end{bmatrix}, C = \begin{bmatrix} c_1 & c_2 & \cdots & c_n \end{bmatrix} ,$$

$$D = \begin{bmatrix} c_o \end{bmatrix} \tag{26}$$

Notice that for $(A, B, C)_N$ system having a finite impulse response, the rows and columns of zeros are omitted and the following finite Hankel matrix is used :

$$H = \begin{bmatrix} c_1 & c_2 & \cdots & 0 & c_N \\ c_2 & c_3 & \cdots & c_N & 0 \\ c_3 & c_4 & \cdots & 0 & 0 \\ & & \vdots & & \\ c_N & 0 & \cdots & 0 & 0 \end{bmatrix} \tag{27}$$

The matrices W_c and W_o, defined by Eqns. (7) and (8) have the same dimension of H and W_c , $W_o \in R_{NxN}$

where R_{NxN} is the set of real matrices.

The H matrix (27) is a symmetric matrix so it can be decomposed (SVD-decomposition) to:

$$H = V \Lambda V^T \tag{28}$$

where

$$VV^T = I \tag{29}$$

and Λ is a diagonal matrix with the eigenvalues of H (Hankel singular values), and I is a unit matrix. Notice that H is not necessary to be a positive-definite matrix.

For the system $(A, B, C)_N$ where A, B, and C matrices are given in Eqn. (26), the controllability matrix, and controllability grammians are unit matrices , i.e.,

$$P = I \tag{30}$$

$$W_c = I \tag{31}$$

and the transformation matrix tool

$$T = V \left| \; \Lambda \; \right|^{1/2} \tag{32}$$

with lead to balanced realization of the system Eqn. (26) where V and Λ are defined by Eqns. (28) and (29), respectively. And, $\left| . \right|$ denotes the absolute value of the matrix elements.

3.3 Balanced truncation:

Considering the fact that the full-order filter of Eqn.(26) is optimal. An optimal reduced one should be close to the optimal one form the input-output of view. Furthermore, it is known from the applied the transformation process on the $(A, B, C)_N$ by (T) according to Eqn. (31) results in a balanced system $(\overline{A}, \overline{B}, \overline{C})_N$.

Next, assume that the matrices Λ, andV can be decomposed each on into two parts as:

$$\Lambda = \begin{bmatrix} \Lambda_1 & 0 \\ 0 & \Lambda_2 \end{bmatrix} \tag{33}$$

where

$$\Lambda_1 = diag(v_1, v_2, \ldots, v_n) \tag{34}$$

and

$$\Lambda_2 = diag(v_{n+1}, v_{n+2}, \ldots, v_N) \tag{35}$$

$$|v_i| = |\sigma_i| \tag{36}$$

where σ_i's are the singular values of F(z)and

$$V = \begin{bmatrix} V_1 & V_2 \end{bmatrix} \tag{37}$$

where V_1, and V_2 are (N x n), and (N x N-n) rectangular matrices, respectively.

According to the partition, the balanced system coordinates can be represented as

$$\overline{A} = \begin{bmatrix} \overline{A}_{11} & \overline{A}_{12} \\ \overline{A}_{21} & \overline{A}_{22} \end{bmatrix}, \ \overline{B} = \begin{bmatrix} \overline{B}_1 \\ \overline{B}_2 \end{bmatrix}, C = \begin{bmatrix} \overline{C}_1 \\ \overline{C}_2 \end{bmatrix} \tag{38}$$

where

$$\overline{A}_{11} = |\Lambda_1|^{1/2} V_1^T A V_1 |\Lambda_1|^{-1/2} \tag{39}$$

$$\overline{A}_{12} = |\Lambda_1|^{1/2} V_1^T A V_2 |\Lambda_2|^{-1/2} \tag{40}$$

$$\overline{A}_{21} = |\Lambda_2|^{1/2} V_2^T A V_1 |\Lambda_1|^{-1/2} \tag{41}$$

$$\overline{A}_{22} = |\Lambda_2|^{1/2} V_2^T A V_2 |\Lambda_2|^{-1/2} \tag{42}$$

$$\overline{B}_1 = |\Lambda_1|^{1/2} V_1^T B \tag{43}$$

$$\overline{B}_2 = |\Lambda_2|^{1/2} V_2^T B \tag{44}$$

and

$$\overline{C}_1 = C V_1 |\Lambda_1|^{-1/2} \tag{45}$$

$$\overline{C}_2 = C V_2 |\Lambda_2|^{-1/2} \tag{46}$$

The orthogonal matrix V achieves that Eqn. (43) is always existed. The full order system (A, B, C)$_N$ is asymptotically stable. Then, the following two-system found to be asymptotically stable and balanced:

(i)   The truncated system - $(\bar{A}_{11}, \bar{B}_1, \bar{C}_1)$.

(ii)   (ii) The rejected system - $(\bar{A}_{22}, \bar{B}_2, \bar{C}_2)$.

If $\tilde{G}(z)$ is a transfer function obtained by truncating the balanced realization of G(z) to the first n states (if σ 's values of the grammian Σ are ordered in a descending manner whereas if σ's order in a sending fashion, the truncated system $\tilde{G}(z)$ obtained from the last n states then

$$\left\| G(z) - \tilde{G}(z) \right\|_H \leq 2tr(\Sigma_2)$$ (47)

The result (47) is presented in [15] where tr (.) is the trace of a matrix.

In inequality (47), the Hankel norm used because it compromises between two conventional-norms: Euclidean norm and the Chebyshev norms.

There is an error in estimation of fading due to the difference between the ideal and approximated one. The impact of this error mirror on the overall channel estimation. This study until now don't take in account certain transmission system. If single user multi-level QAM (MQAM) is taken, the impact of this interpellator appears in the calculation of received signal to noise ratio (SNR).The instantaneous BER is now given by[17]:

$$BER \leq 0.2\, e^{-1.5\,\bar{\gamma}/(M-1)}$$ (48)

where BER , $\bar{\gamma}$ and M are the bit error rate, average approximated SNR and modulation level, respectively.

The spectral efficiency for fixed M is $\log_2$, the number of bits/symbol.

4 Results:

Numerical results are presented to demonstrate the analyses in the preceding sections. Below sample of responses of the approximated IIR estimators for 3, and 6 level truncation with respect to 0 truncation (without truncation). The magnitude and group delay responses of the low truncation level are closed to the responses of FIR one. The impact on the estimation in this case may be neglected. In the case of high level truncation the responses either for magnitude or phase (or group delay is the derivative of the phase function w.r.t. radian frequency) are differ, so the impact be high on the channel fading parameters estimation (magnitude and phase). In order to compensate this effect may be treat it by using efficient channel coding scheme but this will spend portion from the spectral of the channel. Figure 1 and 2 denotes to the exact and approximate magnitude and group delay of 3_{rd} level and 6th level truncation, respectively. The error norm between the two responses is shown in Figure 3.

5 Conclusion

The paper presented a method to reduce the order of interpolator of PSAM. There is a closed-form solution for the observability grammian and Hankel singular values of FIR filters. The balanced realization technique is then applied to find a reduced-order IIR model with a closed-form bound for the infinity norm of the approximation error. The resultant error bound depends on the order of the original IIR filter. Conditions under which the proposed technique leads to a sufficiently small approximation error are given.

REFERENCES

[1]   L. Hanzo, Y.Jos Akhtman, Li Wang, and M. Jiang," MIMO-OFDM for LTE, Wi-Fi and WiMAX," John Wiley & Sons Ltd, 2011.

[2]   J. K. Caves, " An analyses of pilot assisted modulation for Rayleigh fading channels," IEEE Trans. Veh. Technol., vol. 40, No. 4, Nov. 1991.

[3]   X. Cai and G. B. Giannakis, " Error probability minimizing pilots for OFDM with M-PSK modulation over Rayleigh fading channels," IEEE Trans. Veh. Technol., vol. 53, No. 1, Jan. 2001.

[4]   X. Cai and G. B. Giannakis," Adaptive modulation with adaptive pilot symbol assisted estimation and prediction of rapidly fading Channels," Conference on Information Sciences and Systems, The Johns Hopkins University, March 12-14, 2003.

[5]   M. Karami1, A.Olfat and N. C. Beaulieu1, " GeneralizedMIMOtransmit preprocessing using pilot symbolassisted rateless codes" proceeding IEEE Globecom, 2010.

[6]   M. Necker, and F. Sanzi, J. Speidel " An adaptive Wiener-filter for improved channel estimation in mobile OFDM-systems" International Symposium on Signal Processing and Information Technology, IEEE 28 – pp. 213-216, 30 Dec., 2001.

[7]   Rittwik Jana, and Brian D. Hart " Predictor based multiuser and single user detection for time varying frequency selective Rayleigh fading CDMA channels" Vehicular Technology Conference Proceedings, vol.2, P.P 1306-1310 Tokyo, 2000.

[8]   B. C. Moore, " Principal component analysis in linear system: controllability, observability, and model reduction", IEEE Trans. Automat. Contr., Vol., AC-26, pp. 17-32 Feb. 1981.

[9]   B. Beliczynski, I. Kale and G.D. Cain, " Approximation of FIR by IIR digital filters: an algorithm based on balanced model reduction, " IEEE Trans. Signal Processing, Vol. 40, pp.532-542, 1992.

[10] R. C. Gonzalez, and R. E. Woods, " Digital Image Processing, " Prentice HallUpper Saddle River, New Jersey 07458,2002.

[11] M.Rudko, " A note on the approximation of FIR by IIR digital filters: an algorithm based on balanced model reduction" IEEETran.onsignal processing, Vol. 43. No. 1 , Jan. , 1995.

[12] H. K. Kwan and F. Wang, " State-Space approaches for model reduction of FIR digital filters" IEEE.2006.

[13] P. Heydari, and M. Pedram," Model-order reduction using variational balanced truncation with spectral shaping," IEEE Trans. on circuits and systems-I, VOL. 53, NO. 4, April, 2006.

[14] I. Abou-Faycal, M. M´edard, and U. Madhow, "Binary adaptive coded pilot symbol assisted modulation over Rayleigh fading channels without feedback," IEEE Transactions on Communications, vol. 53, no. 6, pp. 1036–1046, June 2005.

[15] V. R. Dehkordi, and A G. Aghdam " Digital filter reduction," World Automation Congress (WAC), Budapest, Hungary, July 24-26,2006.

[16]   P. J. Patil and M. D. Patil, " Model order reduction of high order LTI system using balanced truncation approximation,"Proceedings of International conference on process automation, control, and computing , p.p 1-6, 20-22 July, 2011.

[17]   J. F. Paris, C. A. Torres and J. T. Entrambasaguas, " Impact of channel estimation error on adaptive modulation performance in flat fading," IEEE trans. on communication, Vol. 52, No. 5, P.P 716-720, May, 2004.

Authors:

Hayder J. Al-battat (hayderjawad@ieee.org) has got his MSc in electrical engineering in 2004, from Basra University where he is currently a PhD candidate. His interesting research areas include: wireless communications, microwave engineering, and mobile communications. Mr. Hayder is a member of IEEE.

Haider M. AlSabbagh (haidermaw@ieee.org) was born in 1970, received his Ph.D. degree from school of electronic information and electrical engineering (SEIEE), Shanghai Jiao Tong University in 2008, and his M.S. degree in communications and electronics engineering from Basrah University in 1996. From 1996 to 2002, he worked in Basrah University as a lecturer. Currently, he is an associate professor in Basra University. His research interests include wireless communication, mobile and wireless networks, data communications, information networks, optical communications, on body communications, and antennas design. Dr. Haider is a member of editorial board for several journals and occupies a TPC committee member of many international conferences, also he has been serving as a referee for many international journals: IET-communications, Wiley international journal of communication systems (IJCS), International Journal of Engineering and Industries (IJEI), International Journal of Advancements in Computing Technology (IJACT), Advances in information sciences and services (AISS), Cyber Journals and international conferences, such as: IEEE WCNC 2010 – Networks, ICCAIE 2010, EPC-IQ01, MIC-CCA2009, ICOS2011, ISCI 2011, ISIEA 2011, IET-WSN 2010, ISWTA2011, ICCAIE 2011, RFM 2011. He has awarded CARA foundation for his project's team. Also, he is an TPC member of international conferences: MIC-BEN2011, MIC-CNIT2011, MIC-CSC2011, MIC-WCMC2011, MIC-CPE2011, ICFCN'12, BEIAC 2012, SURSHII'12, APACE2012, ICOS2012. Dr. Haider is a member of IEEE.

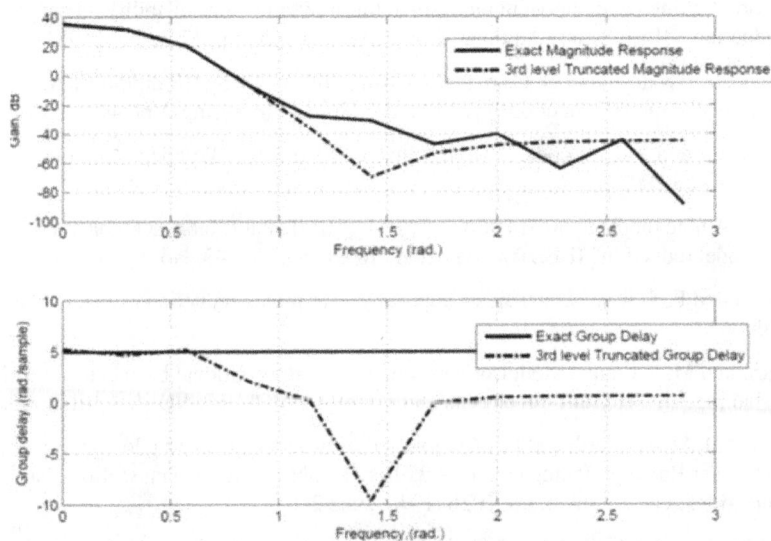

Figure 1: Magnitude response (in dB.) and group delay (in rad./sample) of the approximated IIR estimator for truncation level=3

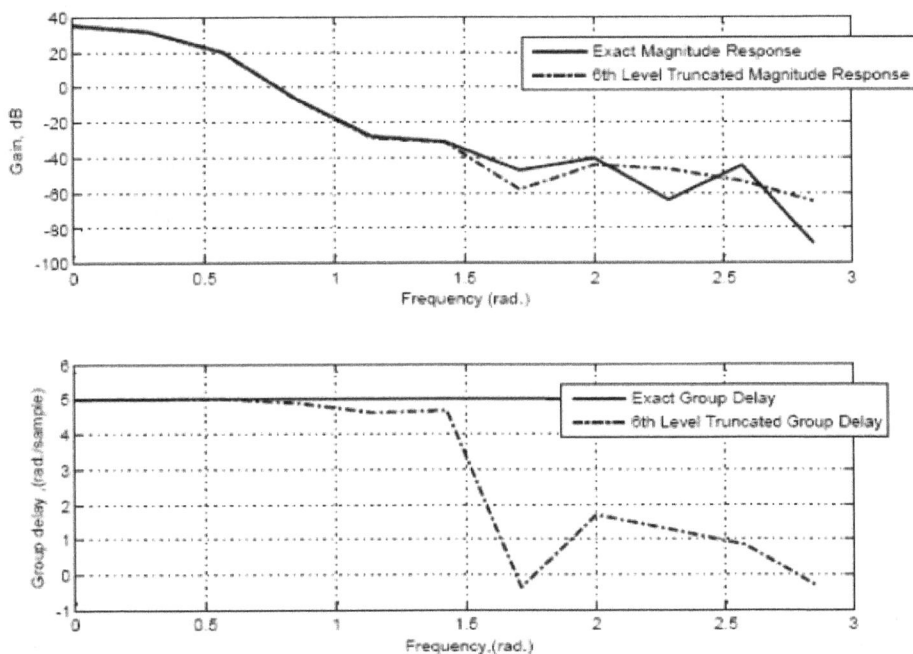

Figure 2: Magnitude response (in dB.) and group delay (in rad./sample) of the approximated IIR estimator for truncation level=6

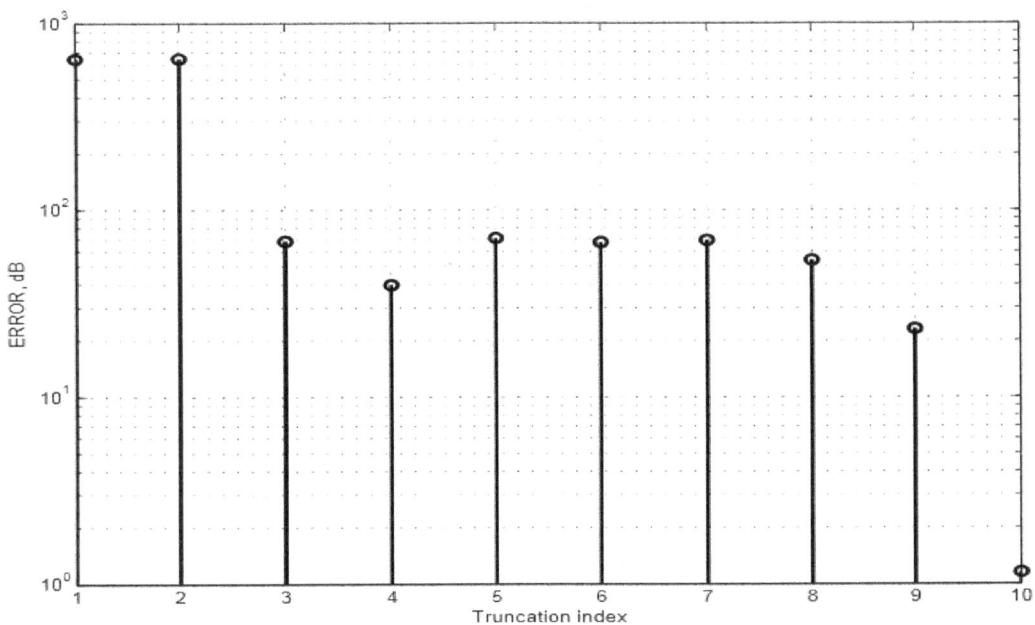

Figure 3: Truncation index versus the error in (dB.)

7

MULTI USER DETECTION FOR CDMA-OFDM/OQAM SYSTEM COMBINED WITH SPACE TIME CODING

Radhia GHARSALLAH[1] and   Ridha BOUALLEGUE[2]

[1]National Engineering School of Tunis, Innov'Com Laboratory, Higher School of Communications, Tunisia
radhia.gharsallah@gmail.com
[2]Innov'Com Laboratory, Higher School of Communications, Tunisia
ridha.bouallegue@gmail.com

ABSTRACT

In this paper, we propose the combination of Multi Carrier (MC) OFDM/OQAM modulation and CDMA-called MC-CDMA-OQAM system with Space Time (ST) coding in a multi user context. This combination takes advantages from multicarrier modulation, spread spectrum and spatial time diversity. Indeed, the use of OFDM has proved its ability to fight against frequency selective channels but the insertion of guard interval yields spectral efficiency loss and sensitivity to frequency dispersion due to the use of rectangular pulse shape. Thus, cyclic prefix OFDM is replaced by an advanced filterbank-based multicarrier system OFDM/OQAM that operates without guard interval. However, OFDM/OQAM provides orthogonality only on the real domain, so transmitted symbols must be real valued. In the other hand, the CDMA component has two advantages: multiple access interference cancellation and providing orthogonality in the complex domain. From the orthogonality property provided, the Alamouti ST code can be combined with MC-CDMA-OQAM system.

The resulting MIMO-MC-CDMA-OQAM system improves the spectral efficiency of wireless system, combat channel fading and reduce narrowband interference. Numerical results show the utility of this new wireless communication system; a significant BER versus Signal to Noise Ratio (SNR) was achieved thus Multiple access interference (MAI) is suppressed and so supporting a large number of users.

KEYWORDS

CDMA, OFDM/OQAM, ST coding, MIMO, Multi User Detection

1. INTRODUCTION

Increasing the transmission rate and providing robustness to channel conditions are nowadays two of the main research topics for wireless communications. Therefore, a significant interest of late has been to develop systems that offer both high capacity and high data speed, along with MAI resistance. Indeed, much effort is done in the area of Multiple Input Multiple Output (MIMO) systems by using several antennas either at the transmitting side or at the receiving side. We can exploit space and time diversity by using Space time codes such as the famous Alamouti code [1]. In multi antennas communication systems, investigations of N_t transmit and N_r receive antennas showed that the capacity of such systems increases linearly with the minimum of N_t and N_r [1], [6]. High data rates are obtained by simultaneously sending signals from several transmit antennas. To protect the integrity of the transmitted information, transmit diversity is obtained by introducing redundancy among the transmitted signals over

N_t transmit antennas (space) and T time periods (time). Under quasi-static fading channel, the maximum combined transmit–receive diversity order equals to $N_t N_r$ [1].

A variety of Space-Time codes exists in the literature. In [6], Alamouti proposed a new modulation scheme over $N_t = 2$ transmit and N_r receive antennas where a rate of one symbol per channel use (PCU) with $2 N_t$ diversity was achieved [8]. The ML detection of the Alamouti scheme can be implemented by a linear complexity decorrelator.   On the other hand, multicarrier modulation (MCM) is becoming the appropriate modulation for transmission over frequency selective channels. The most popular MCM is the Cyclic Prefix Orthogonal frequency Division Multiplexing (CP-OFDM) that exploits frequency diversity by dividing the total bandwidth into M subcarriers and transmitting OFDM symbols over these subcarriers. However, the insertion of the CP yields spectral efficiency loss. In addition, the conventional OFDM modulation is based on a rectangular windowing in the time domain which leads to a poor (sinc(x)) behavior in the frequency domain. Thus CP-OFDM gives rise to two drawbacks: loss of spectral efficiency and sensitivity to frequency dispersion. In order to overcome these two problems, a variant of OFDM that is called OFDM-OQAM has been proposed [3].

OFDM/OQAM overcomes the two drawbacks of OFDM but it does not provide orthogonality on the complex domain, thus the idea is to use CDMA component for two reasons; firstly to avoid multiple access interference and secondly to provide orthogonality in the complex domain if the spreading codes are well chosen [4]. The originality of this paper consists on studying mutli user detection for MIMO-MC-CDMA-OQAM communication system.

This paper is organized as fallows; in section 2 we present the system model of MC-CDMA-OQAM system in order to describe this Multi carrier modulation. In section 3 we will present our new communication scheme combining ST coding with spread spectrum technique and OFDM/OQAM. Section 4 is reserved to ST decoding and multi user detection. In section 5 we present numerical results obtained by matlab simulation showing performance of different multi user receivers of the proposed system. The last section deal with synthesis and comparison between MC-CDMA and MC-CDMA/OQAM in two cases; with and without space time coding and we finish by a conclusion.

2.   MC-CDMA-OQAM SYSTEM MODEL

The block diagram in Figure 1 illustrates the MC-CDMA-OQAM transmission scheme relative to the kth user. It consists on the combination of spread spectrum with multicarrier modulation OQAM [4]. The basic principle of OFDM/OQAM is to divide the transmission bandwidth into a number of subbands. Like for OFDM, the transmitter and receiver can be implemented by using Fast Fourier transform (FFT) algorithms. However instead of a single FFT or IFFT, a uniform filter bank based on a prototype filter is used. At the receiver side the dual operations are carried out.

Different kinds of prototype functions can be implemented such as Isotropic Orthogonal Transform Algorithm or other prototypes optimized in discrete-time using the Time-Frequency Localization criterion.

Let us introduce the CDMA technique used in our system. We denote by Q the spreading factor and K the number of users, each user k is distinguished by its spreading code

$$c_k = \left[c_{0,k}, c_{1,k}, \ldots, c_{Q-1,k} \right]^t$$

Walsh Hadamard codes are chosen because they are characterized by their perfect orthogonality.

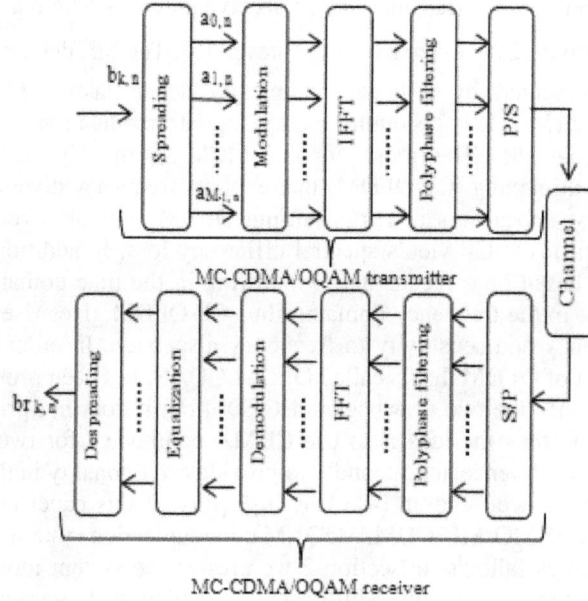

Figure 1 : MC-CDMA-OQAM system model

We consider the transmission of complex data in a free distorsion channel. The spreaded resultant signal of K users can be written as

$$x_{m,n} = \sum_{k=0}^{K} c_k b_{k,m,n} \tag{1}$$

With $b_{k,m,n}$ is the data of the kth user transmitted at time n over the mth subcarrier.

For each user, $b_{k,m,n} = a_{m,n}$

The baseband equivalent of a continous time Multicarrier MC-CDMA-OQAM signal is expressed as follows.

$$s(t) = \sum_{m=0}^{M-1} \sum_{n \in \square} x_{m,n} g(t - n\tau_0) \upsilon_{m,n} e^{j2\pi m F_0 t} \tag{2}$$

With : M the number of subcarrier, $g(t)$ The pulse shape, $F_0 = 1/T_0 = 1/2\tau_0$ the subcarrier spacing and $\upsilon_{m,n}$ an additional phase term.

The length of the prototype filter is a multiple of the number of subcarriers.

We denote by y the received signal, the output of the polyphase filters is obtained as follows.

$$y_{m0,n0} = \langle y, g_{m0,n0} \rangle \tag{3}$$

Then, we apply the despreading block to obtain signal relative to each user k. we get

$$yd_{n0,k} = \sum_{m0=0}^{M-1} c_k y_{m0,n0} \tag{4}$$

The despreaded signal is written as

$$yd_{n0,k0} = b_{n0,k0} + j\left(\sum_{k=0}^{K} \sum_{n=-2b+1}^{2b-1} b_{n+n0,k} \left(\sum_{p=0}^{M-1} \sum_{m=0}^{M-1} c_k c_{k0} \gamma_{m,n+n0}^{(p,n0)} \right) \right) \tag{5}$$

With $\gamma_{m,n}^{(p,n0)} = \Im\left\{ (-1)^{m(n+n0)} j^{m+n-p-n0} A_g(n-n0, m-p) \right\}$

A_g is the ambiguity function of the prototype function g defined as

$$A_g(n,m) = \int_{-\infty}^{\infty} g(u - n\tau_0) g(u) e^{j2\pi F_0 u} du$$

3- SPACE TIME CODING IN MC-CDMA-OQAM SYSTEM

OFDM/OQAM provides orthogonality in the real domain, a pseudo Alamouti code was proposed in [8]. This pseudo code is very complex so we think to take advantage from The CDMA-OFDM/OQAM combination that results the orthogonality in the complex domain so we can combine space time coding to this multi carrier communication system. We consider 2 transmit antennas and 2 receive antennas where are assumed to be far enough apart such the complex fading coefficients among the antennas are uncorrelated. We apply Alamouti ST coding scheme to each user k data.

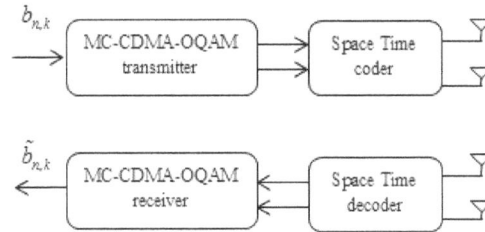

Figure 2 : MC-CDMA/OQAM system with space time coding

We dote by $h_{i,j}$ the complex channel coefficient between the ith transmit antenna and the jth receive antenna. The channel amplitudes are independents, zero mean complex Gaussian variables with unit variance.

The received code word is given by [6] :

$$Y - HX + n \tag{6}$$

With X the transmitted code word, H is channel matrix and n is an Additive White Gaussian Noise.

The channel between transmit antenna i and receive antenna j may be modeled by a complex multiplicative distortion $h_{i,j}(t)$. Assuming that fading is constant across two consecutive symbols, we can write

$$h_{i,j}(t) = h_{i,j} = \alpha_{i,j} e^{j\theta_{i,j}} \tag{7}$$

Form equations (6) and (7), we can write

$$\begin{pmatrix} y_1 & y_3 \\ y_2 & y_4 \end{pmatrix} = \begin{pmatrix} h_{1,1} & h_{1,2} \\ h_{2,1} & h_{2,2} \end{pmatrix} \begin{pmatrix} s_{n_0,k} & -s_{n_0+1,k}{}^* \\ s_{n_0+1,k} & s_{n_0,k}{}^* \end{pmatrix} + \begin{pmatrix} n_1 & n_3 \\ n_2 & n_4 \end{pmatrix} \tag{8}$$

With $s_{n_0,k}$ and $s_{n_0+1,k}$ data of kth user transmitted by the first antenna respectively at time n_0 and n_0+1 after spreading and passing through the polyphase filters.

$-s_{n_0+1,k}^*$ and $s_{n_0,k}^*$ data of kth user transmitted by the second antenna.

y_1 and y_3 signals received by the first antenna,

y_2 and y_4 signals received by the second antenna.

$$\begin{cases} y_1 = y_1(t) = h_{1,1}s_{n_0,k} + h_{1,2}s_{n_0+1,k} + n_1 \\ y_2 = y_2(t) = h_{2,1}s_{n_0,k} + h_{2,2}s_{n_0+1,k} + n_2 \\ y_3 = y_3(t+T) = -h_{1,1}s_{n_0+1,k}^* + h_{1,2}s_{n_0,k}^* + n_3 \\ y_4 = y_4(t+T) = -h_{2,1}s_{n_0+1,k}^* + h_{2,2}s_{n_0,k}^* + n_4 \end{cases} \tag{9}$$

n_1, n_2, n_3, n_4 are complex random variables representing noise.

4. SPACE TIME DECODING AND MUD

4.1. Space time decoding

We consider system with (CSI) Channel State Information at the receiver. Combined signals are given by:

$$\tilde{s}_{n_0,k} = \left(\alpha_{1,1}^2 + \alpha_{1,2}^2 + \alpha_{2,1}^2 + \alpha_{2,2}^2\right)s_{n_0,k} + h_{1,1}n_1 + h_{1,2}n_2 + h_{2,1}n_3 + h_{2,2}n_4 \tag{10}$$

$$\tilde{s}_{n_0+1,k} = \left(\alpha_{1,1}^2 + \alpha_{1,2}^2 + \alpha_{2,1}^2 + \alpha_{2,2}^2\right)s_{n_0+1,k} - h_{1,1}n_2^* + h_{1,2}n_1^* - h_{2,1}n_4^* + h_{2,2}n_3 \tag{11}$$

These combined signals are then sent to the Maximum likelihood (ML) detector. The ML decision rule, at the receiver, is to choose $s_{j,k}$ if and only if

$$\sum_{i=1}^{2K}\sum_{n=1}^{2}d^2\left(y_i, h_{i,n}s_{j,k}\right) \le \sum_{i=1}^{2K}\sum_{n=1}^{2}d^2\left(y_i, h_{i,n}s_{m,k}\right) \forall j \ne m$$

Where $d^2(x,y) = (x-y)(x^*-y^*)$

The output of the ST decoder will be sent to the MC-CDMA-OQAM receiver. At this stage, the first step is to apply polyphase filtering and the second step is to apply multi user detector. The polyphase filtering, at the receiver, consists on using M uniform filters like at the transmitter with a simple FFT.

4.2. Multi user Detection (MUD)

We consider K users communicating simultaneously, each of them transmit L data. Collectively the matched filter output for all K users can be expressed in a long vector as [5]:

$$Z = \left[Z(1)^t \dots Z(L)^t\right]^t \tag{12}$$

With $Z(i) = \left[Z_1(i) \dots Z_K(i)\right]^t$

We denote by $C = [c_1, c_2, \dots, c_K]$ the vector containing spreading codes of K users, $A = I(K)$ the identity matrix of size K and B the vector containing the data of K users. From equation (12), we can write:

$$Z = CC^t AB + n \tag{13}$$

Let w be a linear transformation vector for the multi user detector. The decision vector is

$$d = w^t Z \tag{14}$$

The Zero Forcing (ZF) detector in [2] has a linear transformation equivalent to the inverse of the correlation matrix $R = CC^t$

$$w = R^{-1} \tag{15}$$

The decision vector is then,

$$d = R^{-1}(RAB + n) \tag{16}$$

The decision vector has covariance matrix

$$E\left[(R^{-1}n)(R^{-1}n)^H \right] = \sigma^2 R^{-1} \tag{17}$$

Which can results, in noise power enhancement, creating a gap between the single user error performance and the Zero Forcingr error probability.

Another linear detector with the same structure as ZF detector based on the optimization of the minimum mean-squared error (MMSE) criteria:

$$w = \min\left(E\left[(b - \tilde{b})^t (b - \tilde{b}) \right] \right) \tag{18}$$

The solution of the equation above is

$$w = \left(R + \sigma^2 (A^t A)^{-1} \right)^{-1} \tag{19}$$

While the single user matched filter combats white noise exclusively and the Zero Forcing eliminates Multiple Access Interference (MAI) disregarding background noise, the MMSE linear detector forms a compromise between the two, taking the relevant importance of noise and interfering users into account. Differently to linear MUD there is iterative MUD such Successive Interference Cancellation (SIC) receiver [7] which is less complex.

5. SIMULATION AND NUMERICAL RESULTS

In this section, we present numerical results of the MC-CDMA/OQAM communication system.

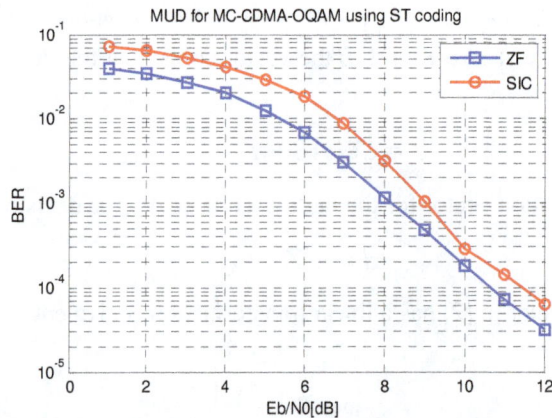

Figure 3: Mutli user detection for Space time coding MC-CDMA/OQAM communication system.

We assume that both the mobile and the base station have two transmit and two receive antennas, we consider K users communicating simultaneously, each user has his spreading code of length Q=16. We use Quadrature Amplitude Modulation QAM 4. The length of fast Fourier Transform is 64.

In Figure 3 multi user detection is presented. It represents Bit Error rate (BER) versus Signal to Noise Ratio (SNR). Simulation results improve that linear multi user detectors provide better performance gain in comparison with nonlinear detectors. However, nonlinear detectors can attend the compromise performance and simplicity.

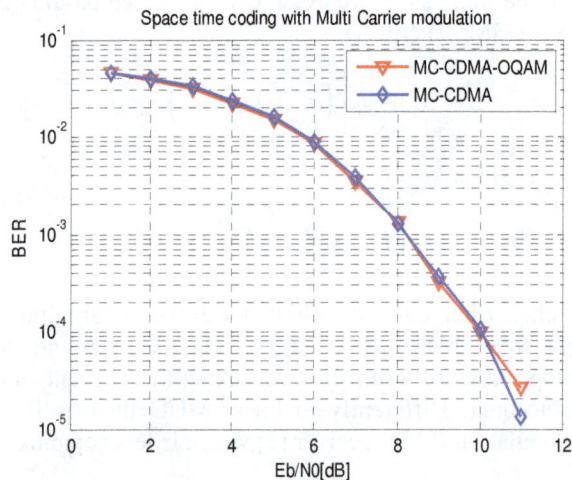

Figure 4: Space time coding with mutli carrier modulation: Comparison between MC-CDMA/OQAM and MC-CDMA.

Figure 4 shows a comparison between the classic MC-CDMA and MC-CDMA/OQAM. MC-CDMA is based on the combination of CDMA and OFDM with cyclic prefix but MC-CDMA/OQAM is based on the combination of CDMA and the advanced multi carrier modulation OFDM/OQAM. Simulation results improve that MC-CDMA/OQAM provide similar BER as MC-CDMA. Furthermore MC-CDMA/OQAM provides better data rate since that operates without guard interval.

In Figure 5 we compare the two multi carrier modulations MC-CDMA and MC-CDMA/OQAM in, both, Single Input Single Output (SISO) and Multiple Input Multiple Output (MIMO) channels. Both SISO and MIMO channels attend good performance in term of BER versus

SNR. In addition, MIMO channel gives the possibility of exploiting the space and time diversity by transmitting different data symbols simultaneously the thing that increases the data rate in comparison with SISO channel.

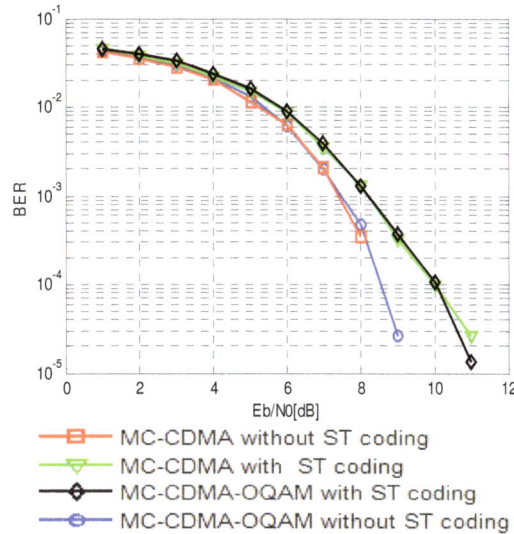

Figure 5: Comparison between MC-CDMA/OQAM and MC-CDMA with and without space time coding.

6. CONCLUSION

In this paper, we have studied and evaluated multi user detection in MC-CDMA-OQAM system in presence of MIMO channel. The idea is to use multicarrier modulation in order to combat channel effects. Two kinds of multi carrier modulations are compared; cyclic prefix OFDM and OFDM/OQAM. OFDM/OQAM operates without guard interval so it eliminates spectral efficiency loss present in OFDM. The idea of combining the CDMA component with multi carrier modulation has two advantages; the first one is to eliminate multiple access interference and the second one is to take advantage from the orthogonality in the complex domain provided by CDMA technique. The resulting is an MC-CDMA/OQAM communication system that provides orthogonality in the complex domain. So we can use space time coding in order to exploit spatial and time diversity. The ST code used is the Alamouti code. Numerical results improve the performance gain of multicarrier modulation in two cases; with and without space time coding in multi user environment. Both linear and iterative multi user detectors applied to the proposed system attend a BER of 10^{-4} for a Signal to Noise Ratio (SNR) approximately equal to 10 dB. As perspective, we think to extend this work to a large number of transmit and receive antennas and to use other ST coders.

REFERENCES

[1] S.M. Alamouti "A Simple Transmit Diversity Technique for Wireless Communications," *IEEE Journal on selected areas in communication, VOL. 16, NO. 8, OCTOBER 1998*

[2] R. Lupas and S. Verdú, "Near -far resistance of multiuser detectors in asynchronous channels," *IEEE Transaction in communications., vol. 38, pp. 496 -508, Apr.1990.*

[3] H. Bolcskei, "Orthogonal frequency division multiplexing based on offset QAM," *in Advances in Gabor Analysis, Birkhauser, Boston, Mass, USA, 2003.*

[4]   C. Lélé, P. Siohan, R. Legouable, and M. Bellanger, "CDMA transmission with complex OFDM/OQAM," *EURASIP Journal on Wireless Communications and Networking, vol. 2008, Article ID 748063, 12 pages, 2008.*

[5]   P. Patel and J. M. Holtzman, "Analysis of a simple successive interference cancellation scheme in a DS-CDMA system," *IEEE Journal on selected areas in communications., vol. 2, pp. 796-807, June 1994.*

[6]   H. Huang, H. Viswanathan, and G. J. Foschini, "Multiple Antennas in Cellular CDMA Systems: Transmission, Detection, and Spectral Efficiency," *IEEE Transactions on Wireless Communications., vol. 1, pp.383 -392, July 2002.*

[7]   T. Muharemovic, E. N. Onggosanusi, A. G. Dabak, and B. Aazhang, "Hybrid linear-iterative detection algorithms for MIMO CDMA systems in multipath channels," *in Proc. IEEE Int. Conf. Acoustics, Speech, and Signal processing, 2002 (ICASSP '02), vol. 3 , pp. 2621 -2624, 17 May 2002.*

[8]   H. Lin, C. Lélé, and P. Siohan, "A pseudo alamouti transceiver design for OFDM/OQAM modulation with cyclic prefix," in *Proceedings of the IEEE Workshop on Signal Processing Advances in Wireless Communications (SPAWC '09)*, pp. 300–304, Perugia, Italy, June 2009.

8

LDPC DECODER MODELLING AND EVALUATION USING RT-PEPA

Tony Tsang[1]

[1]Hong Kong Polytechnic University, Hung Hom, Hong Kong.
ttsang@ieee.org

ABSTRACT

This paper presents a high-throughput memory efficient decoder for Low Density Parity Check (LDPC) codes in the high-rate wireless personal area network application. The novel techniques which can apply to our selected LDPC code is proposed, including parallel blocked layered decoding architecture and simplification of the WiGig networks. We use Real Time - Performance Evaluation Process Algebra (RT-PEPA) to evaluate a typical LDPC Decoder system's performance. The approach is more convenient, flexible, and lower cost than the former simulation method which needs develop special hardware and software tools. Moreover, we can easily analysis how changes in performance depend on changes in a particular modes by supplying ranges for parameter values.

KEYWORDS

LPDC, IEEE 802.15.3c, RT-PEPA, Performance Analysis, Formal Modelling.

1. INTRODUCTION

Low density parity check (LDPC) codes, first proposed by Gallager in 1962 have attracted much attention because of their excellent error correcting performance, inherently parallelism and high throughput potentials. Therefore, they are being widely used in communication standards such as IEEE 802.16e and IEEE 802.11n. In addition, millimeter wave (mmWave) Wireless Personal Area Networks (WPANs) described by the IEEE 802.15.3c Working Group are considering LDPC codes as the preferred choice for forward error correction (FEC).

Recently, many studies have been accomplished to simplify the VLSI implementation of the related decoders called "Architecture-Aware LDPC codes" or "Block-LDPC codes". Based on these design approaches, quasi-cyclic LDPC (QCLDPC) codes have received significant attentions due to their efficient hardware implementations. Furthermore, QC-LDPC codes can provide comparable error-correction performance compared with random LDPC codes.

With the increasing demand for high-data-rate wireless application, an overlapped decoding architecture and a layered decoding architecture are popular for high-throughput LDPC decoding architectures using QC-LDPC codes. An overlapped decoding architecture based on the partially parallel architectures using the belief-propagation (BP) algorithm, where the message updating computations in the check node unit (CNU) and in the variable node unit (VNU) are partially overlapped, can increase the decoding throughput by maximizing the hardware utilization efficiency without any performance degradation.

Compared with the overlapped two-phase decoding scheme, a layered decoding algorithm and architecture have been proposed to achieve a faster convergence. Generally, layered decoding algorithm is the horizontal layered decoding algorithm which is favorable for the Min-Sum algorithm. The layered decoding algorithm offers 2x throughput and significant memory advantages, compared with the BP algorithm. Also, it can reduce the average number of iteration using intermediate check-node (or variable-node) message values. However, conventional layered decoders use a bidirectional network or two switch networks for shuffling and reshuffling messages, which increases the hardware complexity. Also, due to the data dependency between consecutive rows in layered decoding, the parallel and pipelining techniques cannot be applied directly. In this paper, we develop a fully pipelined architecture targeted for the IEEE WiGig standard that has fully parallelized variable nodes and layer serialized check nodes. By exploiting the structure of the lower rate codes, it reduces the number of sub-iterations by up to half.

The rest of the paper is structured as follows. Section II introduces Real-Time Performance Evaluation Process Algebra (RT-PEPA) and its Tools. Section III describes the design choices made for the architecture of Decoder Functional Blocks Specification, and Section IV details performance evaluation result of the functional design of the decoder's major blocks. Then, conclusion is presented in Section V.

2. RT-PEPA and its Tools

We present a formal modelling language, called Real-Time Performance Evaluation Process Algebra (RT-PEPA), to describe the real-time stochastic behaviour of communication systems. The language combines conventional stochastic process algebra with real-time semantics to describe complex systems in a compositional manner. It includes timed transition, parallel composition, probabilistic branching and hard real-time aspects. Performance Evaluation Process Algebra (PEPA), developed by Hillston in the 1994s [3, 4], is a timed and stochastic extension of classical process algebras such as Communication Sequential Process (CSP) [5]. It describes a system as an interaction of the components and these components engage in activities. Generally, components model the physical or logical elements of a system and activities characterize the behavior of these components. An exponentially distributed random variable is associated with each activity specifies the duration of it, that leads to a clear relationship between the model and a Continuous Time Markov Chain (CTMC) process. Via this underlying Continuous Time Markov Chain process performance measures can be extracted from the model. The PEPA formalism provides a small set of timed operators which are able to express the individual real time activities of components as well as the interactions between them. We provide a brief summary of the operators here, more details about PEPA can be found in [2, 3].

2.1. Notation and Definitions of RT-PEPA

Real-Time - Performance Evaluation Process Algebra (RT-PEPA) is process algebraic language which supports the compositional description of concurrent and distributed systems and analysis of their performances. The basic elements of RT-PEPA are its actions, which represent activities carried out by the systems being modeled, and its operators, which are used to compose algebraic descriptions.

Time point

A time point is a time instant with respect to the global clock of the real time system; it does not have duration. It specifies the starting and stopping times of an action. Using a time point, we can instruct the system to generate an action at a particular point in time. Time point progresses consistently in all parts of the system. More formally, the time point is defined by using a discrete time domain, which contains the following properties:

$$\forall t \, \exists t' \, t < t' \wedge \forall t'' : t < t'' \Rightarrow t' \leq t''$$

We assume a fixed set of clock $t = \{t_0, \ldots, t_i\}$. The special time point $t0$, which is called the start time point, always has the value 0.

Time Constraint

An action can exist for a short period of time; this duration is called the time constraint of the action. A time constraint has a starting and an ending point. It consists of a lower-bound and an upper-bound time point, where the lower-bound time point enables an action in a module, and the upper-bound time point disables the action at that point in time. Formally, we define a time constraint in the following:

$$\mathrm{T}_i = \{ [\tau_{i_{min}}, \tau_{i_{max}}] \ \forall t_i \in T \} \ with \ 0 \le \tau_{i_{min}} \le \tau_{i_{max}}$$

Timed Action

A timed action is a tuple $< \alpha, \lambda, \mathrm{T} >$ consisting of the type of the action α, the rate of the action λ and temporal constraint of the action T. The type denotes the kind of action, such as transmission of data packets, while the rate indicates the speed at which the action occurs from the view of an external observer. The rates are used to denote the random variables specifying the duration of the actions. The actions can be defined in different types of probability distribution function such as Exponential, Poisson, Constant, Geometric and Uniform distribution. Moreover, each transition is also bounded by a temporal constraint. In this section, some basic notations and operation semantics about RT-PEPA are briefly introduced. The syntax of RT-PEPA is defined in the following:

$$P ::= stop \mid < \alpha, \lambda, \mathrm{T} > . P \mid P + Q \mid P \oplus_{r,\mathrm{T}} Q \mid P \otimes_{L,\mathrm{T}} Q \mid P \ \square_{P,\mathrm{T}} Q \mid P / L \mid A$$

The conventional stochastic process algebra operators and the additional operations are described in the following:

- *stop* is an inactive process

- $< \alpha, \lambda, \mathrm{T} > . P$, which stands for a prefix operator, where the type of the action is a probability distribution function (pdf) type α, with the activity rate denoted by λ, and the temporal constraint of component is T. It subsequently behaves as P. Sequences of actions can be combined to build up a time constraint for an action. The time constraint T is defined as above.

- $P + Q$ is choice combinator capturing the possibility of competition or selection between different possible activities. It represents a system which may behave either as P or as Q processes. All the current actions P and Q process are enabled. The first action to complete distinguishes one of the processes. The other process of the choice is discarded. The system will then behave as the derivative resulting from the evolution of the chosen process.

- $P \oplus_{r,\mathrm{T}} Q$ denotes the probabilistic choice with the conventional generative interpretation, thus with probability r the process behaves like P and with probability $1 - r$ it behaves like Q bounded with the time constraint T.

- $P \otimes_{L,\mathrm{T}} Q$ is a cooperation, in which the two actions P and Q are parallel, synchronizing on all activities whose type is in the cooperation set L of action types. The lifetime of two actions is the time constraint T. These two actions are disabled when the time constraint expires. Any action whose type is not in L will proceed independently. As a syntactic convenience the parallel combinator is defined by $\otimes_{\emptyset,\mathrm{T}}$, where the cooperation set L is empty and the lifetime of two actions is T.

- $P\ \square_{\text{P, T}}\ Q$ is a unary operator which returns the set of actions that meet the temporal predicate condition specified by T . P consists of several predicates combined with the boolean connectives: `And' ,`Or', Exclusive-Or (EXOR)' and `Not'. $\square_{\text{And, T}}$ means both actions can occur during the interval T . $\square_{\text{or, T}}$ means that one or both actions can occur during the interval T . $\square_{\text{EXOR, T}}$ means that one of these actions occurs; it immediately determines whether P or Q can subsequently occur during the triggered interval T . $\square_{\text{Not, T}}$ means that both actions do not occur during the interval T .

- $P\ /\ L$ is a hiding operation, where the set L of visible action types identifies those activities which are to be considered internal or private to the component. These activities are not visible to an external observer, nor are they accessible to other components for cooperation.

- $A := P$ is a countable set of constants.

2.2. RT-PEPA Eclipse Plug-in Tools

The PEPA is a language for modelling systems in which a number of interacting components run in parallel, and whose behaviour is stochastic. The core semantics of PEPA is in terms of Continuous Time Markov Chains (CTMCs), and an alternative semantics in terms of Ordinary Differential Equations (ODEs) has also been developed. PEPA has been applied in practice to a wide variety of systems, and its success as a modelling language has been largely down to its extensive tool support. Most recently, the PEPA Plug-in Project [6, 7] has integrated a range of analysis techniques based on both numerical solution and simulation into a single tool built on top of the Eclipse platform [8]. As with all compositional Markovian formalisms, however, PEPA suffers from the state space explosion problem. A model can have an underlying state space that is exponentially larger than its description, meaning that it can be infeasible to analyse. Fluid flow approximation using PEPAs ODE semantics can solve this problem if we are only interested in the average behaviour of the system over time. However, if we want to reason over all possible behaviours of the model for example, the probability that an error occurs within some time interval then we must consider the CTMC semantics. In this paper, we present a new extension to the PEPA plugin, in which a model can be abstracted by combining, or aggregating, states. To safely over-approximate the behaviour of the original model (for any aggregation of its states), we use two abstraction techniques - abstract CTMCs (a type of Markov decision process with infinite branching), and stochastic bounds. We provide a model checker for the three-valued Continuous Stochastic Logic (CSL), which computes from the abstraction a safe bound of the probability of a quantitative property holding in the original model X if the actual probability is p, then the model checker will return an interval I = [p1, p2] such that p 2 I. The current version of the PEPA plug-in is available from http://www.dcs.ed.ac.uk/pepa/tools/plugin, and provides several views:

2.2.1 Abstract Syntax Tree View

The Abstraction View is a graphical interface that shows the state space of each sequential component in a PEPA model. It provides a facility for labelling states (so that they can be referred to in CSL properties), and for specifying which states to aggregate.

2.2.2 Model Checking View

The Model Checking View is an interface for constructing, editing, and model checking CSL properties. The property editor provides a simple way to construct CSL formulae, by referencing the labels given to states in the abstraction view. It ensures that only syntactically well-formed CSL formulae can be constructed.

2.2.3 State Space View

The State Space View is linked to the active PEPA editor and provides a tabular representation of the state space of the underlying Markov chain. The table is populated automatically when the state space exploration is invoked from the corresponding top level menu item. A row represents a state of the Markov chain, each cell in the table showing the local state of a sequential component. The state space order in which sequential components are displayed corresponds to the order in which they are found in the co-operation set by depth-first visit of the co-operations binary tree. A further column displays the steady-state probability distribution if one is available. A toolbar menu item provides access to the user interface for managing state space filters. When a set of filter rules is activated, the excluded states are removed from the table. The probability mass of the states that match the filters is automatically computed and shown in the view. Filter rules are assigned names and made persistent across workspace sessions. From the toolbar the user can invoke a wizard dialogue box to export the transition system and one to import the steady-state probability distribution as computed by external tools. The view also has a Single-step Debugger, a tool for navigating the transition system of the Markov chain. The debugger can be opened from any state of the chain and its layout is as follows. In an external window are displayed the state description of the current state and two tables. The tables show the set of states for which there is a transition to or from the current state. The tables are laid out similarly to the views main table. In addition, the action types that label a transition are shown in a further column. The user can navigate backwards and forwards by selecting any of the states listed.

2.2.4 Performance Evaluation View and Graph View

Performance Evaluation View and Graph View A wizard dialogue box accessible from the top-level menu bar guides the user through the process of performing steady-state analysis on the Markov chain. The user can choose between an array of iterative solvers and tune their parameters as needed. Performance metrics are calculated automatically and displayed in the Performance Evaluation View. It has three tabs showing the results of the aforementioned reward structures (throughput, utilisation, and population levels). Throughput and population levels are arranged in a tabular fashion, whereas utilisation is shown in a two-level tree. Each top-level node corresponds to a sequential component and its children are its local states. The Performance Evaluation View can feed input to the Graph View, a general purpose view available in the plug-in for visualising charts. Throughputs and population levels are shown as bar charts and a top-level node of the utilisation tree is shown as a pie chart. As with any kind of graph displayed in the view, a number of converting options is available. The graph can be exported to PDF or SVG and the underlying data can be extracted into a comma separated value text file.

2.2.5 Experimenting with Markovian Analysis

An important stage in performance modelling is sensitivity analysis, i.e. the study of the impact that certain parameters have on the performance of the system. A wizard dialogue box is available in the plug-in to assist the user with the set-up of sensitivity analysis experiments over the models. The parameters that can be subjected to this analysis are the rate definitions and number of replications of the array of processes in the system equation. The performance metrics that can be analysed are throughput, utilisation, or population levels. If the model has filter rules defined, the probability mass of the set of filtered states can be used as a performance index as well. The tool allows the set-up of multiple experiments of two kinds: one-dimensional (performance metric vs. one parameter) or two-dimensional (performance metric vs. two parameters changed simultaneously). The results of the analysis are shown in the Graph View as line charts. For example, a parameter that may have an important impact on the performance of the real time system is the reset delay of the CPU.

2.2.6 Time Series Analysis

When performing a time-series analysis there are three basic steps to complete; component selection, solver selection and solver parameterization, all of which are handled by the time-series analysis wizard. Rather than simply observing all components, the wizard allows the modeller to select only those components that are of interest. This becomes more pertinent as either the number components in the system or number of observed time points increase - one limitation of the current time-series solvers is that all data is held in memory, and only written out to disk when exporting from the graph view. Solver selection and parameterization are self-explanatory, with the list of visible parameters being dynamically linked to the currently selected solver. In keeping with the rest of the UI, the selections across all three steps are persistent across invocations. Likewise, each unique parameter is stored only once, meaning parameters such as start and stop times are persistent over all solvers. Lastly, the parameters, including selected solver, are attached to the results in the graph view for future reference. Currently this meta data can only be seen when the data is exported. The last feature of the wizard is the ability to export the model in alternative formats, such as Matlab.

3. Decoder Functional Blocks Specification

3.1. Overall System

The decoder has five major blocks: variable nodes (VN), check nodes (CN), barrel shifters (BS), pre-routers, and post-routers. Figure 1 shows the high-level connection of all the blocks. The VNs are combined into 16 groups of 42VNs, called a variable node group (VNG). Each VN within a VNG connects to one port of a 42-input barrel shifter to implement the sub-matrix shifts. The outputs of the barrel shifter are further routed by the pre-routers, which connect to one of the 16 inputs of the 42 CNs. The outputs of each CN go through inverse shifting using post-routers and another set of barrel shifters. The design has several levels of hierarchy in order to keep irregular wiring local and make the global wires as regular as possible. Since the design layer serializes the CNs, they are time multiplexed to act as different CNs in each cycle. The CN has all of the information it needs to compute the new check to variable (C2V) message from all CNs in the first layer, and, after it has finished processing the inputs, it sends back a single message to all neighboring VNs. In the next cycle, the VNs send another single message that has been marginalized for the CNs in the second layer, so that the same CNs can compute the C2V message from the CNs of the second layer. This continues for all the layers in the matrix, with serial messages being passed back and forth from VNs to CNs. The decoder uses the flooding schedule because layered decoding has too many dependencies to be used effectively in a highly parallel, fully pipelined design. It processes one layer per pipeline stage, and the VNs accumulate one frames messages while sending out the others. The decoder equation defines how the components interact with each other. According to the working cycle and the definitions of model's components we give before, the decoder equation is show below:

$$Decoder_0 := Back - Shifter_0 \otimes_{L,T} Variable - Node_0 \otimes_{L,T} Front - Shifter_0 \otimes_{L,T}$$
$$Pre - Router_0 \otimes_{L,T} Check - Node_0 \otimes_{L,T} Decoder_0 ;$$

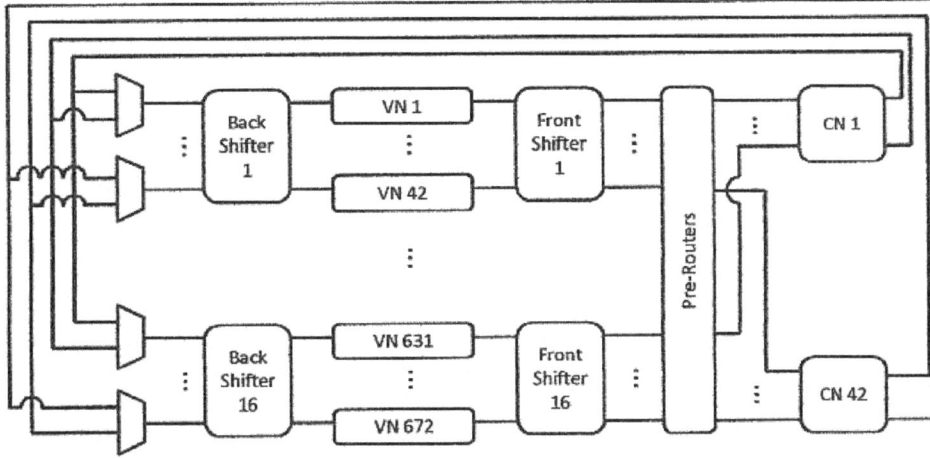

Fig. 1. Decoder Block Diagram

3.2. Variable Node

The VN implements [9] and performs both the CN and VN marginalization, which keeps all memory within the VN. When starting to decode a new data frame, the VN loads in a prior value to either the first or second frames prior or accumulation registers. Over the next four cycles sends variable to check (V2C) messages to its neighboring CNs. Since the VN performs the C2V marginalization, it locally stores the V2C messages it just sent out in a shift register. Once the un-marginalized C2V message returns, the VN compares the first minimum magnitude to the V2C magnitude from the shift register. If they match, the VN uses the second minimum; otherwise, it uses the first minimum. It marginalizes the sign by taking the XOR of the new and stored V2C signs. The VN accumulates the marginalized C2Vs, with the prior value added in when the first C2V arrives, and it also stores them individually in another shift register. After all C2V messages have been accumulated, the VN calculates the new V2C messages by subtracting out the saved C2V values from the accumulated value, which is the V2C marginalization. Also, the sign bit of the accumulated value is used as the hard decision and for early termination. Figure 2 shows the block diagram for the VN.

$$Variable - Node_0 := \beta Corrector_0 \otimes_{L,T} S / M - 2'sComp_0 \otimes_{L,T} XOR_0 \otimes_{L,T}$$
$$Accum_0 \otimes_{L,T} Subtractor_0 \otimes_{L,T} 2'sComp - S / M_0 ;$$

$$\beta Corrector_0 := <listen, \lambda_0, T_0 > .C2V - Min_0 \otimes_{L_0} < get, \lambda_0, T_0 > .Mux_0 ;$$

$$Subtractor_0 := <listen, \lambda_0, T_0 > .Mux_0 \otimes_{L_0} < get, \lambda_0, T_0 > .V2C - Marg_0 ;$$

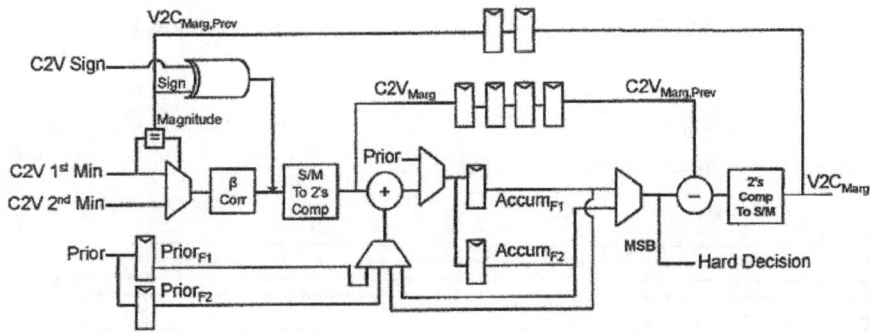

Fig. 2. Variable Node Schematic

3.3. Check Node Node

Since the VN performs the marginalization, the CN only needs to find the first and second minima of all the incoming V2C message magnitudes and the product of the V2Cs signs. The CN computes the former with a compare select tree. A sorting block at the beginning arranges pairs of inputs in ascending order, and subsequent stages of the tree select the minimum two out of four inputs. A separate XOR tree finds the product of the signs. The CN is modified to process either a single high-weight layer or two non-overlapping layers. To process a single layer, it takes the output of the entire tree to find the first and second minima of all 16 inputs. For non-overlapping layers, the weight is at most 8, so the top half of a tree can process one layer and the bottom half can process the other. The outputs are taken at the second to last stage to obtain first and second minima for each layer. This partitions a 16-input CN into two 8-input CNs, giving the CN two levels of granularity. Because the layers do not overlap, no read before write conflicts occur. Figure 3 shows the final CN architecture for the magnitude computation. The sign's XOR tree is partitioned similarly.

$$Check-Node_0 := Sort_0 \otimes_{L,T} CS_0 \otimes_{L,T} CS_1 \otimes_{L,T} XOR_0 \otimes_{L,T} CS_2 \ ;$$

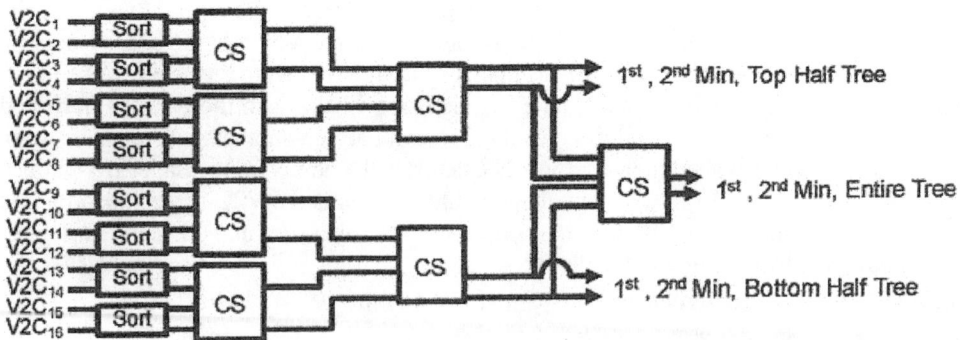

Fig. 3. Check Node Magnitude Computation Schematic

3.4. Pre- and Post-Routing

When processing two non-overlapping layers at once, barrel shifters alone cannot guarantee that the first layers inputs will go to the top half of each CN and the second layers inputs will go to the bottom half. Granular CNs need two extra sets of routing to allow this, as depicted in Figure 4. Pre-routing comes before the CN and selects which VNGs go to the top half or bottom half of each CN. Post-routing comes after the CN and, for each VNG, selects whether to send the top trees or bottom trees result. Both types of routers are implemented with a small number of muxes. In the case of processing one layer, the routers do not shift the messages.

$$Router_0 := Select_0 \otimes_{L,T} Pre-Router_0 \otimes_{L,T} Check-Node_1 \otimes_{L,T} Post-Router_0 \otimes_{L,T} Select_1 \, ;$$

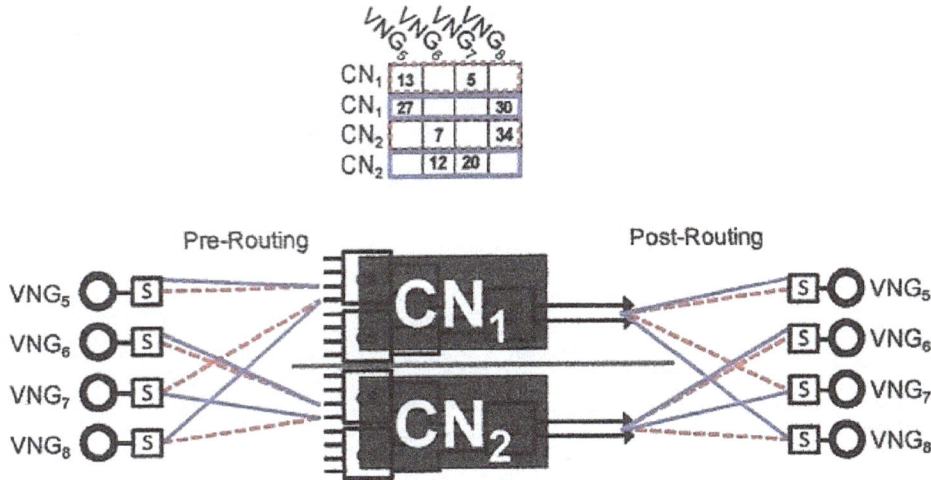

Fig. 4. Operation of Pre- and Post-Routers

3.5. Pipeline

The decoder has five pipeline stages and processes two independent data frames simultaneously. Each full iteration takes three sub-iterations for the rate 13/16 code and four for the rest, one for each time-multiplexed use of the check nodes. In the first stage, the VN outputs the marginalized V2C and the barrel shifter reorders the messages. The second stage consists of the pre-routing and global wiring to the check node. The CNs processes their inputs in the third stage and route the messages back to the VNs across the global wires in the fourth stage. In the fifth stage, the VN accumulates the serial messages over three or four cycles. The accumulation would normally cause a four cycle bubble in the pipeline, it due to the dependency between accumulating all the C2Vs and sending out the next V2Cs. The bubble is just large enough to accommodate processing a second frame. This reduces the bubble to one cycle for the rate 13/16 code and removes it completely for the other rates because no dependencies exist between the two frames. Figure 5 gives the pipeline diagram for the decoder with four sub-iterations.

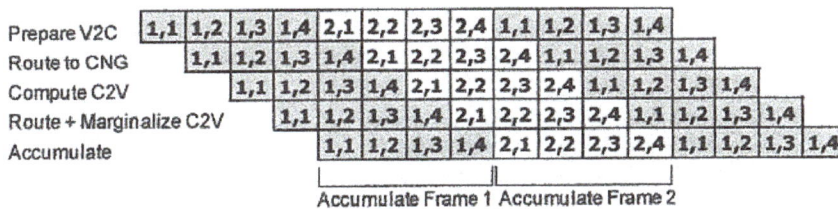

Fig. 5. Pipeline Diagram for rate 1/2, 5/8, and 3/4 codes where the first number indicates the frame and the second the sub-iteration

4. PERFORMANCE EVALUATION

The study of how changes in performance depend on changes in parameter mode values is known as sensitivity analysis. We can vary some parameter's value a little, and see its influence degree to the model's performance, for example, the throughput or response time. Throughput is an action-related metric showing the rate at which an action is performed at steady-state. In other words, the throughput represents the average number of the activities completed by the system during one unit time. From Figure 6, it can be observed that the impact of the number of devices on the throughput of transmit is more sensitive than the QPSK-1/2 and 16-QAM-1/2 modulation for NLOS channel CM 2.3 and LOS channel CM 1.2 Modes. If we could make some efforts to optimize the cache, and raise the 16-QAM-1/2 modulation for NLOS channel Mode form 0.6 to 0.85 or even more high value, the throughput of transmit could greatly improved.

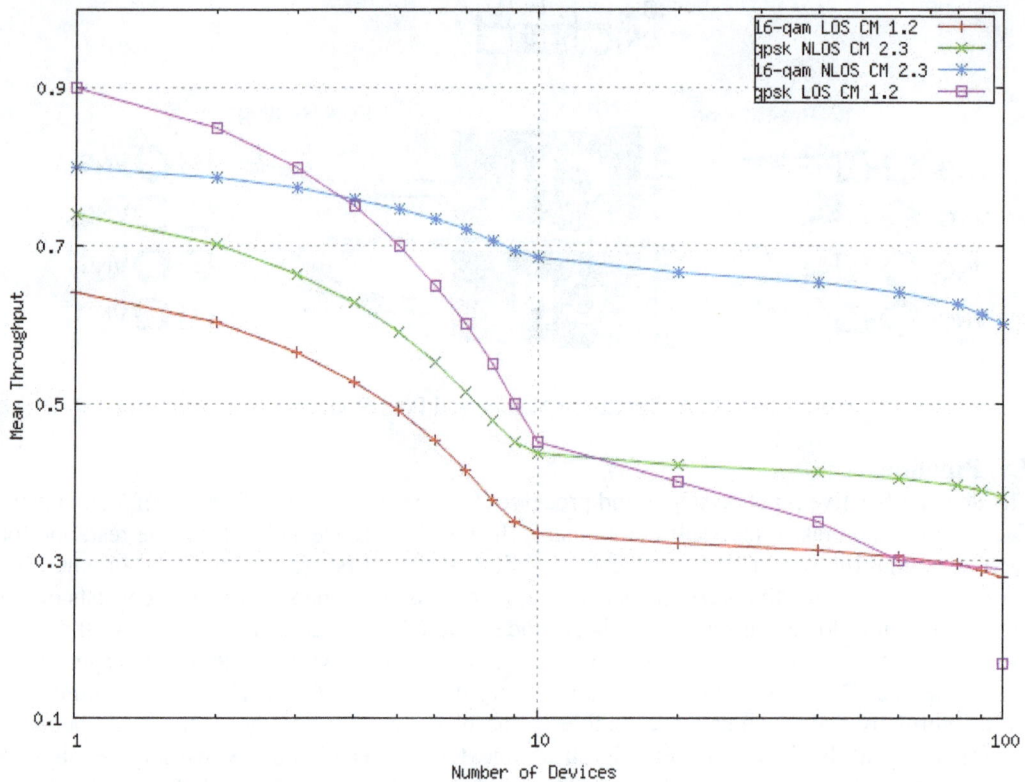

Fig. 6. Throughput versus Number of Devices

For channel coding, we have investigated the performance of LDPC (768,384) coded OFDM systems All receiver functions including packet detection, fine timing synchronization, channel estimation equalization and common phase-error correction have been included. Quantization of bit metrics has been performed with 5 soft-bits in both coding schemes. The chosen LDPC decoding algorithm is the min-sum algorithm with a maximum of 25 iterations per LDPC block. The performance for QPSK-1/2 and 16-QAM-1/2 modulation for NLOS channel CM 2.3 and LOS channel CM 1.2 are presented in Figure 7. The simulation results show better performance of LDPC code for both channel models. The target for the Bit Error Rate (BER) ranges from 10% down to 1%. In the presence of NLOS channel model (CM) 2.3, LDPC coding achieves higher coding gain, in the range of 1.6 to 2.7 dB for QPSK, and 1.0 to 2.5 dB for 16-QAM. On the other hand, in the presence of LOS channel CM 1.2, the coding gain of the LDPC code is approximately 1 dB.

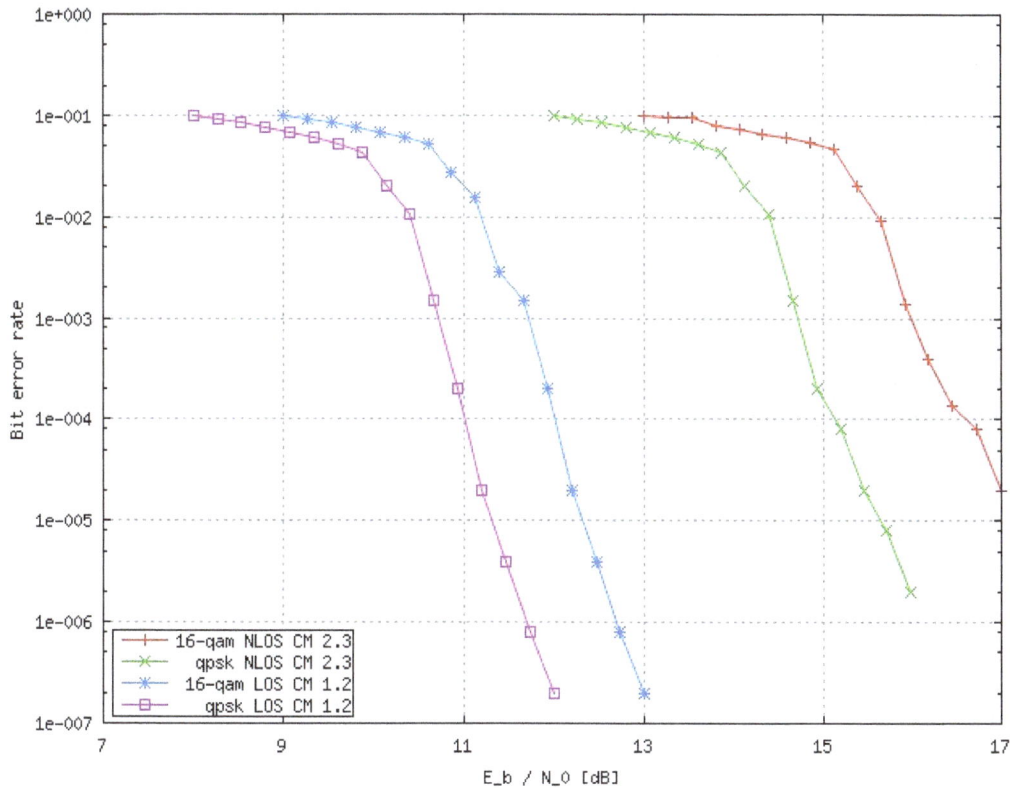

Fig. 7. BER of designed LDPC Decoder for 60-GHz NLOS Channel CM2.3 & LOS channel CM 1.2

5. CONCLUSIONS

The proposed fully pipelined architecture with granular check nodes supports all matrices and low-power modes of the IEEE 802.11ad standard. The routing complexity was reduced significantly by replacing a crossbar based interconnect network with a fixed wire network for SN. Hence, the proposed decoder architecture has high throughput, low interconnect complexity and very low decoding latency. The architecture can be extended to other short block length codes that have the same property of non-overlapping layers, such as to be incorporated in next-generation high-rate WPAN applications.

REFERENCES

[1]. IEEE Std 802.15.3c-2009 (Amendment to IEEE Std 802.15.3-2003), "IEEE Standard for Information Technology - Telecommunications and Information Exchange between Systems - Local and Metropolitan Area Networks - Specific Requirements. Part 15.3: The ultimate purpose of the 60-GHz WPAN systems is to deliver MAC throughput of the order of multi-Gb/s over a reasonable range. To accomplish this, system designers have to increase the transmission range, especially in non-line-of-sight channels. IEEE Communications Magazine E July 2011 121 Wireless Medium Access Control (MAC) and Physical Layer (PHY) Specifications for High Rate

Wireless Personal Area Networks (WPANs) Amendment 2: Millimeter - Wave-Based Alternative Physical Layer Extension", pps. c1-187, Octaber. 2009.

[2].   Tony Tsang, "Performance modelling and evaluation of OFDMA based WiMAX systems using RT-SPA", Proceedings of the International Conference on Computer and Communication Engineering 2008 (ICCCE08), Kuala Lumpur, Malaysia, pps. 180- 186, May 13-18, 2008.

[3].   J. Hillston, "A Compositional Approach to Performance Modelling", PhD Thesis, The University of Edinburgh, 1994.

[4].   J. Hillston, "Fluid flow approximation of PEP A models", Proceedings of the Second International Conference on the Quatitative Evaluation of Systems, IEEE Computer Society Press, pps. 33-41, 2005.

[5].   C.A.R.Hoare, "Communicating Sequential Process", Prentice-Hall, 1985.

[6].   Micheal J.A. Smith, "Abstraction and Model Checking in the PEPA plug-in for Eclipse", Seventh International Conference on the Quantitative Evaluation of Systems, pps. 155-156, 2010.

[7].   M. Tribastone, A. Duguid, and S. Gilmore. "The PEPA Eclipse Plug-in", Performance Evaluation Review, 36(4):28-33, March 2009.

[8].   The Eclipse platform. http://www.eclipse.org.

[9].   Weiner Matthew; Nikolic Borivoje; Zhang Zhengya; "LDPC decoder architecture for high-data rate personal-area networks", IEEE International Symposium on Circuits and Systems (ISCAS), pps. 1784 - 1787, 15-18 May, 2011.

9

MOBILE ELEMENTS SCHEDULING FOR PERIODIC SENSOR APPLICATIONS

Bassam A.alqaralleh and Khaled Almi'ani

Al-Hussein Bin Talal University

ABSTRACT

In this paper, we investigate the problem of designing the mobile elements tours such that the length of each tour is below a per-determined length and the depth of the multi-hop routing trees bounded by k. The path of the mobile element is designed to visit subset of the nodes (cache points). These cache points store other nodes data. To address this problem, we propose two heuristic-based solutions. Our solutions take into consideration the distribution of the nodes during the establishment of the tour. The results of our experiments indicate that our schemes significantly outperforms the best comparable scheme in the literature.

I. INTRODUCTION

Many typical applications of wireless sensor networks (WSNs) considers the process of data collection. This data collection is usually accomplished by wireless transmission of the data (possibly) through multiple hops when sensors are deployed in a hostile or hard-to-access environments. In many cases, energy efficiency has been a major concern in wireless communications because increasing energy expenditure limits the operational lifetime of the network. Furthermore, the exhaustion of the energy sources of the sensors in multi-hop scenarios is non-uniform, as nodes that are close to the sink carry heavier data traffic loads and therefore they are likely to be the first to run out of energy. Once these sensors fail, the sink nodes cannot be reached, and as a result, the network stops working even though the nodes that are located far away from the sink may still have sufficient energy. This is a common problem regardless of which communication protocols are used in the network.

In general, in order to significantly increase the lifetime of the network, Mobile Elements (MEs) [1], [2], [3] have been used because it roams in the network and collects data from sensors via short range communications, therefore, the energy consumption is considerably reduced. Thus, the lifetime of the network increases by avoiding multi-hop communication. The main disadvantage of using this approach is the increased latency of the data collection because the speed of mobile element is typically about $0.1 - 2\ m/s$ [4] [5], which results in extensive traveling time for the ME and, also, delay in gathering the data from sensors.

In practice, the ME tour length is often bounded by a predetermined time deadline, either due to timeliness constraints on the sensor data or due to the limited amount of energy available to the ME itself. A possible solution to this problem is to employ more than one ME; however, this solution is often impractical since the cost of MEs is high, and also it might be useless when some sensors are beyond reach of MEs due to its battery limitations.

To solve this problem, many proposals presented a hybrid approach which has been proposed to be a combination of using multi-hop forwarding and the use of mobile element(s). In this approach, a mobile element visits subset of the nodes that is selected to be caching points. These caching points store the data of the nodes that are not included in the tour of the mobile element. Each caching point transmits its data to the mobile element when it becomes within its transmission range. By adopting such an approach, the mobile elements will be able to collect the data of the entire network without the need for visiting each node physically. Figure 1 shows an example for this hybrid approach, where the mobile element visits the caching points to collect the data of the entire network.

In this direction, we investigate the problem of designing the tour of the mobile element and the data forwarding trees, with the objective of minimizing the depth of forwarding trees. We propose two heuristic-based solutions to address this problem. The first heuristic works by recursively partitioning the network; based on the distribution of the nodes. Then in each partition, the process works to determine the caching points that satisfies the constraints. The second heuristic employs similar steps on tree-structured network. The results of our experiments indicate that our schemes significantly outperforms the best comparable scheme in the literature.

The rest of the paper is organized as follows. Section 2 presents the related work in this research area. Section 3 presents the Problem definition. In Section 4, we present an Integer linear program formulation for the presented problem. In Section 5, we present the details of our algorithmic solutions. In section 6, we extended our heuristics to address the situation, where more than one mobile element is available. Section 7 presents the evaluation. Finally, Section 8 concludes the paper.

II. RELATED WORK

This section reviews the recent literature that studied the use of mobile element(s) to extend the lifetime of sensor networks. We review three major approaches based on the categorization given in [3].

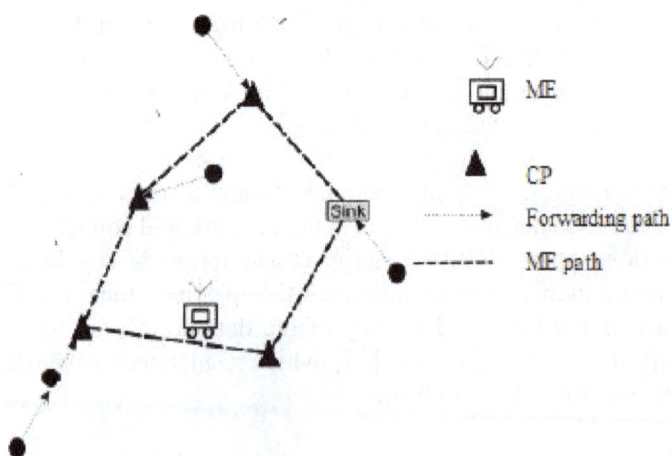

Figure 1: An example showing the mobile element path and the forwarding trees

Firstly, in a typical flat-topology network, the nodes around the sink suffers from heavy load of forwarding the data traffic from all other sensors, consequently, these nodes are likely to be the first to die. Several proposals [6], [7] have investigated the use of mobile sink(s) to reduce the energy consumption in the network. Using this technique, the remaining energy in the nodes becomes more evenly balanced throughout the network by varying the path to the sink(s), leading to a longer lifetime of the network. However, this technique requires the routes to sink node and its location to be changed regularly, which may potentially cause excessive overhead at the nodes due to the frequent re-computation of the routes. Zhao et al [8], [9] presented two distributed algorithms to maximize the overall network utility. These data gathering algorithms are based on placing the mobile sink at each anchor point (gathering point) for a certain period of sojourn time and, on the other hand, mobile sink collects data from nearby sensors via multi-hop communications. They considered both cases of fixed and variable sojourn time.

In the second approach, the paths of the mobile elements are designed to visit each node. In this approach the data of each node is gathered by a mobile element via single-hop communication. In this situation, the path planning problem share fundamental characteristics with the Traveling Salesman Problem (TSP) [10]. It is clear that new constraints must added to capture the characteristics of the sensor environments. In [11], [12], [13], [14], [15], [16], [17], [18], [19] proposed several heuristics to visit each node. This approach provides the capability of significantly reducing the energy consumption by avoiding multi-hop communications, however, it incurs a high delay when the network area is large because the MEs must physically visit all sensor nodes.

Finally, the third approach is a hybrid approach that combines data collection by mobile elements with multi-hop forwarding .Our work is along the line of this approach. Some previous proposals, e.g. [20], [21], [22], assumed that mobile route should be predetermined, and also, were mainly concerned with the timing of transmissions in order to minimize the need for in network caching via timing the transmissions to coincide with the passing of the tour. The problem presented in this share some similarities with minimum-energy Rendezvous Planning Problem (RPP) [23], [24]. In this problem, the objective is to determine the mobile element path such the total Euclidean distance between nodes not included in the tour and the tour is minimized. In [24], the authors presented the Rendezvous Design for Variable Tracks (RD-VT) algorithm. The process of this algorithm starts by construction the Steiner Minimum Tree (SMT) that connects the source nodes. Then, the obtained tree will be traversed in pre-order until no more nodes can be visited without violating the deadline constraint. In this algorithm the visited nodes is identified as the caching points. Xing et al. [23] provided a utility-based algorithm and address the optimal case for restricted version of the problem. Many proposals [25], [26], [27] have also investigated this problem. The problem presented in this work can be categorized as a restricted version of the RPP problem, where the main difference is bounding the depth of the routing trees.

The problem presented in this work also share some similarities with the problem presented by Almi'ani et al. [28]. Their problem deal with designing multiple connected tours such that each node is most k-hope away from one of the tours. The main difference between this problem and our problem is that we focus on the single mobile element case. Also, in the multiple mobile situation, we do not restrict the tours to be connected.

The problem presented in this paper shares some similarities with the Vehicle Routing Problem (VRP) [29]. Given a fleet of vehicles assigned to a depot, VRP deals with the determination of the fleet routes to deliver goods from a depot to customers while minimizing the vehicles' total travel cost.

III. PROBLEM DEFINITION

We are given an undirected graph $G = (V, E)$, where V is the set of vertices representing the locations of the sensors in the network, and E is the set of edges that represents the communication network topology, i.e. $(v_i, v_j) \in E$ if and only if v_i and v_j are within each others communication range. The complete graph $G' = (V, E')$, where $E' = V \times V$, represents the possible movements of the mobile elements. Each edge $(v_i, v_j) \in E'$ has a length $r_{i,j}$, which represents the time needed by a mobile element to travel between sensor v_i and v_j . The data of all sensors must be uploaded to a mobile element periodically at least once in L time units, where L is determined from the application requirements and the sensors buffer size. In other words, we assume that each mobile element conducts its tour periodically, with L being a constraint on the maximum tour length. In this paper, for simplicity, we assume that the mobile element travels at constant speed, and that, therefore, the travelling times between sensors ($r_{i,j}$) correspond directly to their respective Euclidean distances; however, this assumption is not essential to our algorithms and can be easily dropped if necessary. Also, we are given k that represent the maximum number of hops allowed between any node and its caching point.

In our problem, we seek to find the mobile element tour, where the length of the tour is bounded by L, such that the depth of the multi-hop routing trees is minimized and bounded by k.

IV. INTEGER LINEAR PROGRAM FORMULATION

In this section, we present an Integer Linear Program (ILP) for the investigated problem. This ILP is based on the formulation proposed by Almi'ani et al. [26]. We modify their formulation by incorporating the constraint that restrict the number of hops between any node and its caching point to k, as an upper bound.

Variables

$y_{i,j}$, $y_{i,j} = 1$ if the edge (v_i, v_j) is included in the ME tour, and 0 otherwise
$x_{i,j}$, $x_{i,j} = 1$ if the node v_i is included in the tour and is responsible for storing the data of node v_j , which is not included in the tour. $x_{i,j} = 0$ otherwise

Parameters

$d(v_i, v_j)$ is the number of hops between v_i and v_j .
$r_{i,j}$ is the travelling time between v_i and v_j .

Objective

$$\min \sum_{(i,j)} y_{i,j} \cdot r_{i,j} + M_l \cdot \sum_{(i,j)} x_{i,j} \cdot d(v_i, v_j) \tag{1}$$

The first term of the objective function is the travelling time of the ME (travel cost), and the second term is the number of hops between the nodes not included in the tour and the tour (assignment cost). The coefficient M_l must be set to any value greater than $max_{i,j} r_{i,j}$. Multiplying the second term by M_l ensures that the assignment cost is strictly prioritized over the travel cost.

Constraints

• The load at each node is balanced

$$\sum_{i \in V} y_{i,j} - \sum_{i \in V} y_{j,i} = 0 \qquad\qquad \forall j \in V, \tag{2}$$

- the tour starts and ends at vs.

$$\sum_{i \in V} y_{v_s,i} = 1 \tag{3}$$

$$\sum_{i \in V} y_{i,v_s} = 1 \tag{4}$$

- bounds the travelling time of the ME tour.

$$\sum_{i,j \in V} y_{i,j} \cdot r_{i,j} \leq L \tag{5}$$

- Each node must be either involved in the mobile element tour or connected to a node involved in this tour

$$x_{i,j} + \sum_{l \in V} y_{i,l} \leq 1 \qquad\qquad \forall i,j \in V \tag{6}$$

$$\sum_{j \in V} y_{i,j} + \sum_{j \in V} x_{i,j} > 0 \qquad\qquad \forall i \in V \tag{7}$$

- the subtour elimination constraint.

$$z_i - z_j + n \cdot y_{i,j} \leq n - 1 \qquad\qquad \forall i \in V, \forall j \in V - v_s \tag{8}$$

- the number of hops between any node and its caching point is less than or equal to k

$$x_{i,j} \cdot d(v_i, v_j) \leq k \qquad\qquad \forall i,j \in V \tag{9}$$

Typically, it is hard to solve such problem. However, the ILP is given to take close look at the problem.

V. THE ALGORITHMIC SOLUTION

We propose two complementary approaches to address the presented problem. In the first approach, we begin by identifying the best nodes to be used as caching points. Once these nodes are identified, the TSP tour that consists of these nodes is constructed. The last step of this approach is to build the forwarding trees rooted at the caching nodes.

In the second approach, by transforming the network topology into tree-structure, we start by identifying the forwarding trees, and then we determine the caching points. However the first approach leaves more room for optimization for the cache point identification and tour finding steps.

As we will present in this section, both heuristics combine approximation algorithms for two fundamental NP-complete problems: The Dominating Set Problem and The Traveling Salesman Tour Problem. Each of these algorithms is used in different phases of the two heuristics, but in different stages of the algorithm. The design of the heuristics presented in the work is inspired by the P-Based algorithm proposed by Almi'ani et al. [28]. However, as we mentioned before, Almi'ani et al. [28] investigated different problem. Firstly, we describe how these phases work.

A. Selecting the Caching points

Selecting the caching points is the first step in the process of constructing the mobile element tour. The problem we are considering requires that all nodes not included in the tour to be within k hops from their caching points. Therefore, the nodes on the tour will be an Extended Dominating Set of the graph. The nodes outside the tour will need to forward their traffic using multi-hop routing trees that as we know will be bounded in size.

B. Constructing the tour

The tour construction step build the mobile element tour to pass through the caching points identified in the previous step. The objective of the tour building step is to minimize the traveling cost, and therefore it is exactly the TSP problem. Any TSP algorithm or heuristic can be used to obtain the tour of the mobile element. Here, we use the Christofides approximation algorithm, as it is known to behave well in practice.

C. Building the routing trees

Nodes outside the tour will need to forward their data using a multi-hop routing tree. As we will present next, both of the presented heuristics need to solve the k-hop dominating set problem that can be defined as follows:

Given a graph $G = (V, E)$, find a subset of the nodes, such that every node in a graph G is at most k hops away from a node in that subset. The subset is called a k-hop dominating set, and we would like to minimize the size of it. To solve the k-hop dominating set problem, we use the algorithm proposed by Almi'ani et al. [28].

To solve the k-hop dominating set, we first need to construct the graph G_k, that will has an edge between any two node, if there is a path between these two nodes in G with at most k-hops. The graph G_k has the same set of vertices as the graph G. Now the process iterate until all nodes removed from G_k. In each iteration, the node with highest degree (the nodes with highest number of edges) and all of its neighbors will be removed. These removed nodes will be considered as one set. Then, the graph G_k will be altered by removing the edges between the removed node and the nodes that is still in G_k. Algorithm 1 shows the steps of this algorithm.

Next we present the Graph Partitioning (GP) and the Tree Partitioning (TP) heuristics.

Algorithm 1 DOMINATING SET APPROXIMATION

1: **procedure** KDOMAPPROX(G,k) $\triangleright$ $G = (V,E)$
2:    Initialize empty sets D, C
3:    Initialize G_k
4:    **while** G_k has nodes **do**
5:        Find $v \in G_k$ that has the highest degree
6:        Remove v and all of its neighbors from G_k
7:        Add to the removed nodes to C
8:        Add C to D
9:        Empty C
10:   **end while**
11:   **return** D
12: **end procedure**

Algorithm 2 THE GRAPH PARTITIONING HEURISTIC

1: **procedure** GB(G,k, L)
2:    Initialize empty sets T, R and CPs
3:    $k = 1$
4:    $T = L + 1$
5:    **while** $T > L$ **do**
6:        $Cps =$ FindCPS(G,k)
7:        $T =$ BuildTour(CPs)
8:        $R =$ BuildRouting(CPs, G)
9:        $k = k+1$
10:   **end while**
11:   **return** CPs,T and R
12: **end procedure**

VI. THE GRAPH PARTITIONING HEURISTIC

The goal of this heuristic is to determine the caching points, such that the length of the path between any node and its caching point is within k-hops. In addition, the caching points are selected with the objective of maximizing the lifetime of the network. Once these caching points are obtained, the process proceeds to build the multi-hop routing trees and the mobile element tour.

As we discussed in the previous section, this heuristic consists of three steps, (1) the caching point identification step, (2) the routing trees construction step and (3) the tour building step. At the beginning the k value will be set to one. Once the mobile element tour is obtained, if the obtained tour violates the transit constraint, the process will be repeated and the value of k will be increased by one. Otherwise, the obtained solution will be confirmed as a valid solution. Algorithm 2 shows the steps of this heuristic.

A. Selecting the Caching points

This step starts by solving the k-hop dominating set problem, using the algorithm proposed in the previous section. Solving the k-hop dominating set problem results in group of sets, where in each set there is a node such that the path between this node and any other nodes in the set is below k-hops. Once these sets are identified, the process proceeds to select the candidate caching points from each set. In each set, a node is selected as a candidate caching point, if the path between this node and any other node in the set is at most k-hops.

Then, the process iterates to confirm the final caching points list. In each iteration, the nearest caching point candidate to the confirmed caching points will be selected as a caching point. At the beginning the confirmed caching point list will contain only the sink node. This selected caching point will be added to the confirmed list. All other candidates belong to the same set as the confirmed caching point will be removed from the consideration. The process stops, when each candidate caching point is either confirmed as a caching point or removed from consideration. Algorithm 3 shows the process of this step.

B. The routing trees construction and the tour building steps

Once the caching points set are identified, each node not included in this set will be assigned to its nearest caching point. Then, for each caching point and the nodes assigned to this caching point, a Minimum Spanning Tree (MST) is created to establish the multi-hop forwarding trees.

Algorithm 3 THE CACHING POINTS IDENTIFICATION STEP

```
1:  procedure FINDCPS(G,k)
2:      Initialize empty sets C, D and CPs
3:      D=kDomApprox(G,k)
4:      for each node v in D do
5:          if path between v and all nodes in its sets with k-hops then Add v to C
6:          end if
7:          add sink to CPs
8:      end for
9:      while C has nodes do
10:         Find the closest node in C to any node in CPs
11:         Move this nodes to CPs
12:         Remove any node that belong to same set as this node from C
13:     end while
14:     return CPs,
15: end procedure
```

As we mentioned earlier, the last step is to create the tour of the mobile elements to involve the caching points and the sink. This is established using Christofides algorithm.

VII. THE TREE PARTITIONING HEURISTIC

The main steps of the tree partitioning heuristic are similar to the steps employed by the graph partitioning heuristic. However, the order in which these steps are used is different in both algorithms.

The tree partitioning heuristic aims to construct its solution by recursively partitioning the routing tree, to identify the caching points and construct the mobile element tour. This heuristic works by first constructing the MST (rooted at the sink) that connect all nodes. By constructing such a tree, the process aims to eliminate costly edges from consideration during the process of obtaining the solution. Then the heuristic proceeds to the caching points identification step. This step is exactly the same as the one in the graph partitioning heuristic. However, the only difference is the type of input, in the TP heuristic, this step uses the MST as an input, where in the GB heuristic, the topology graph is the input. Once the caching points are identified, the routing trees construction and the tour building steps works in the same way as in the GP heuristic. Similar to the GP heuristic, in each iteration, the value of k is incremented by one until a solution that satisfy the transit constraint is obtained.

Algorithm 4 THE CLUSTERING STEP

1: **procedure** FINDCLUSTERS(G (topology graph), c (number of clusters to be established))
2:      Randomly choose c nodes as initial cluster centers
3:      **while** c is not the same in two consecutive iterations **do**
4:          **for** each node v in G **do**
5:              assign each node to the nearest cluster center
6:              assign recalculate center nodes
7:          **end for**
8:      **end while**
9:      **return** *Clusters*,
10: **end procedure**

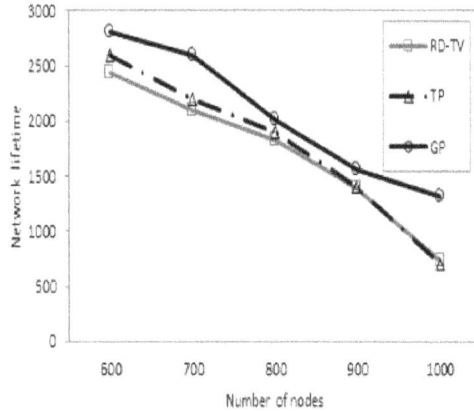

Figure 2: network lifetime against the number of nodes, for the uniform density deployment scenario

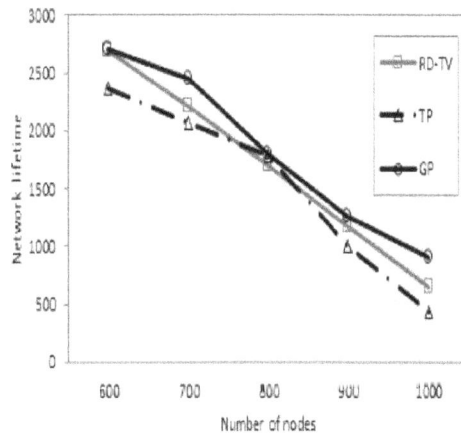

Figure 3: network lifetime against the number of nodes, for the variable density deployment scenario

VIII. MULTIPLE MOBILE ELEMENTS

In this section, we extended our model to address the situation where more than one mobile element is available (M mobile elements), and therefore we need to design the tours for these mobile element. We assume that the number of available mobile elements is not enough to visit each node.

When multiple mobile elements is available, the main design criteria is how to partition the network based on the available number of mobile elements. The nodes distribution must be considered during the mobile element assignment to ensure that the resultant partitioning will obtain an efficient solution. To this end, we propose the Multi- partitioning (MP) heuristic.

The MP heuristic works by partitioning the network into M partitions, where each partition will be assigned to a mobile element. Then in each partition, the GP heuristic is used to obtain the tour mobile element assigned to this partition.

To obtain the mobile elements partitions, the clustering step is used to obtain M number of clusters. In this step, we use the clustering algorithm proposed by Almi'ani et al. [26]. In this algorithm, the objective is to determine the clusters such that the hop-distances between nodes belong to the same cluster is minimized. This aim to balance the number of nodes belong to the same cluster. Toward this end, the process works by bounding the distance between any node and the center node of this cluster. The center node is defined as the node that is the closest to all other nodes inside the cluster.

Algorithm 4 shows the process of this step. The process starts by selecting c number of nodes randomly as the initial center nodes. Then, each node is assigned to its nearest cluster center node. Once all nodes are assigned, the center node for each cluster is recalculated. The nodes then will be assigned to the new center nodes. The process terminates when the identity of the center nodes does not change between two consecutive iterations.

IX. SIMULATION METHODOLOGY AND RESULTS

To validate the performance of the presented algorithms, we have conducted an extensive set of experiments using the J-sim simulator for wireless sensor networks [30]. Unless mentioned otherwise, the network area is $250,000 m^2$. The value of the tour length constraint L is set to $0.15 \cdot s \cdot T_L$, where $s = 1 m/s$ is the speed of the mobile element, and T_L is the length of the minimum spanning tree that connects all nodes. The radio parameters are set according to the MICAz data sheet [31], namely: the radio bandwidth is 250 Kbps, the transmission power is $21 mW$, the receiving power is $15 mW$, and the initial battery power is 10 Joules. For simplicity, we only account for the radio receiving and transmitting energy. Each node generates one packet in an interval of time equal to L. The packet has a fixed size of 100 bytes. Each experiment is an average of 10 different random topologies. We are particularly interested in investigating the following metrics; (1)The lifetime of the network, and (2)Number of caching points. The parameters we consider in our experiments look at varying the number of nodes. We consider the following deployment scenarios:

Uniform density deployment: in this scenario, we assume that the nodes are uniformly deployed in a square area of $500 \cdot 500 \ m^2$.

Variable density deployment: in this scenario, we divide the network into a $10 \cdot 10$ grid of squares, where each square is $500 \cdot 500 \ m^2$. We randomly choose 30 of the squares, and in each one of those we fix the node density to be x times the density in the remaining squares. x is a density parameter, which in most experiments (unless mentioned otherwise) is set to $x = 5$.

As a benchmark to the presented algorithms, we compare their performance against the Rendezvous Design for Variable Tracks (RD-VT) algorithm [24]. The RD-VT algorithm works by constructing the Steiner Minimum Tree (SMT) of the source nodes and build a tour based on this tree. To ensure the fairness of the comparisons, we use the Christofides algorithm to find the TSP tour for a given set of nodes in every iteration of the RD-VT algorithm as well. Eventually, each sensor is connected to the nearest point of the tour via the shortest path. With regard to the comparison with the MP heuristic, we adopt the clustering step and run the RD-VT algorithm in each partition.

For simplicity, we only account for the radio receiving and transmitting energy. Figures 2 and 3 show the results for both deployment scenarios as a function of the number of nodes (equivalently, network density). From the figures we can see the GP heuristic outperforms the TP and the RD-VT heuristics in both deployment scenarios. Also, in the uniform deployment scenario, it is clearly shown that the TP heuristic slightly achieve better performance compared to the RD-VT heuristic. In contrast, in the variable deployment scenario, the RD-VT heuristic outperforms the TP heuristic. To understand the factors behind the shown performances, we need to take a close look on the mechanism of each heuristic. In the RD-VT heuristic, the main idea is to traverse the SMT in pre-order, until no more nodes can be visited without violating the transit constraint. The use of such mechanism results in selecting relatively large number of caching points. However, the selected caching points are expected to be very close to each other and the sink. This expected to results in relatively very deep routing trees, since the caching points are not distributed to cover the entire network. The depth of such routing trees must degrade the performance of the RD-VT heuristic, since it is expected to generate a large amount of forwarding traffic. In the TP heuristic, the use of the k-hop dominating algorithm to partition the MST results in introducing dependence between the TP performance and the structure of the MST. The first partition is expected to have the sink node, since it is the node that normally has the highest number of neighbors. Such mechanism is expected to partition the MST into many branches. This is obvious because after the selection of each dominating set the number of tree-branches is expected to increase. The impact of such mechanism is expected to degrade the TP heuristic, especially in the variable deployment scenario, since in this case the MST is expected to have many branches to begin with. In the GP heuristic, the partitioning occurs based on the distribution of the nodes, and therefore this heuristic is expected to have the same performance ratio, regardless of distribution pattern. These factors clarify the shown performances behavior.

Now, we investigate the impact of the number of nodes on the number of caching points that each heuristic obtains. Figures 4 and 5 show the results for both deployment scenarios. From the figures we can see that in both deployment scenarios, the RD-VT heuristic obtains the highest number of caching points. Also, we can see that the GP heuristic consistently obtains the lowest number of caching points. As we mentioned, the RD-VT heuristic results on selecting caching points very close to each other in term of distance. This is the main key behind the RD-VT heuristic capability of obtaining a high number of caching points. In the GP heuristic, the partitioning step has more control over the distribution of the caching points compared to the TP heuristic. This is clearly due to the number of cuts to the original MST tree, after each partitioning step, in the TP heuristic.

Now, we move to compare the performance of the MP heuristic against the modified version of the RD-VT heuristic. Figures 6 and 7 show the impact of the number of nodes on the lifetime of the networks; for both deployment scenarios. From the figures we can see that in both deployment scenarios, the MP heuristic consistently outperforms the RD-VT heuristic. This is expected, since both heuristics use to same strategy to assign the mobile elements to networks partitions, and therefore this comparison must behave similar to one between the GP and the RD-VT heuristics. Figures 8 and 9 show the impact of number of nodes on the number of caching points each algorithm obtains; for both deployment scenarios. Also, this experiment behavior is similar to the one mobile element experiment because of the same mentioned reasons.

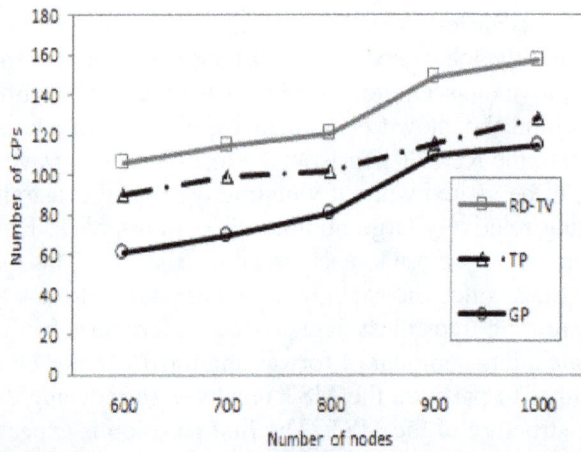

Figure 4: Number of caching points against the number of nodes, for the uniform density deployment scenario

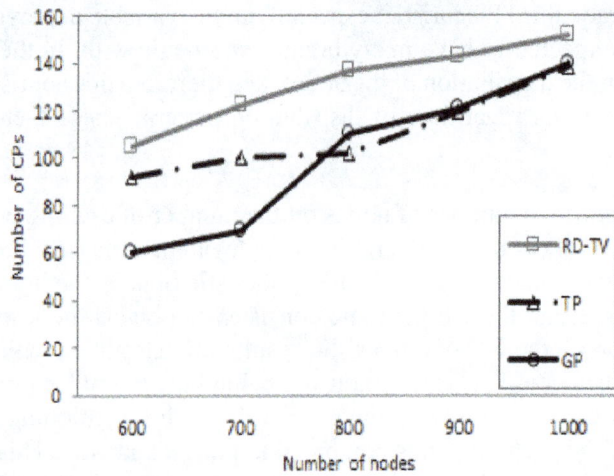

Figure 5: Number of caching points against the number of nodes, for the variable density deployment scenario

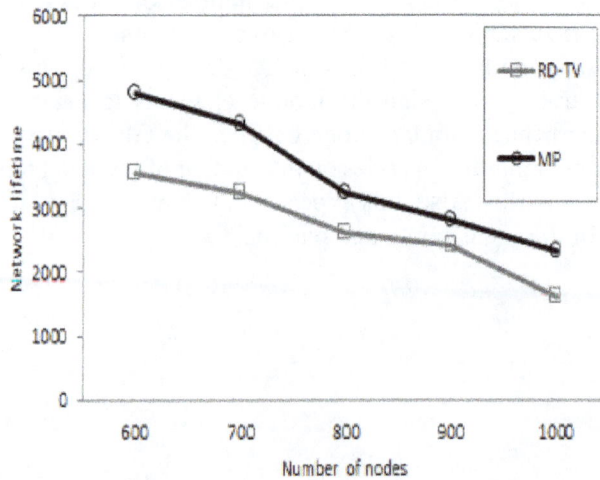

Figure 6: network lifetime against the number of nodes, for the uniform density deployment scenario

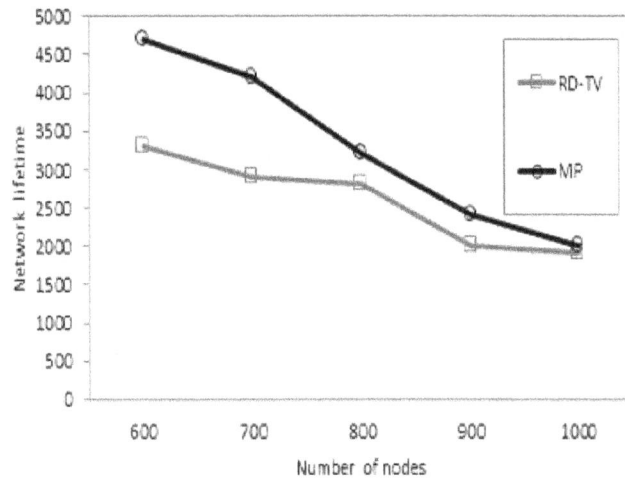

Figure 7: network lifetime against the number of nodes, for the variable density deployment scenario

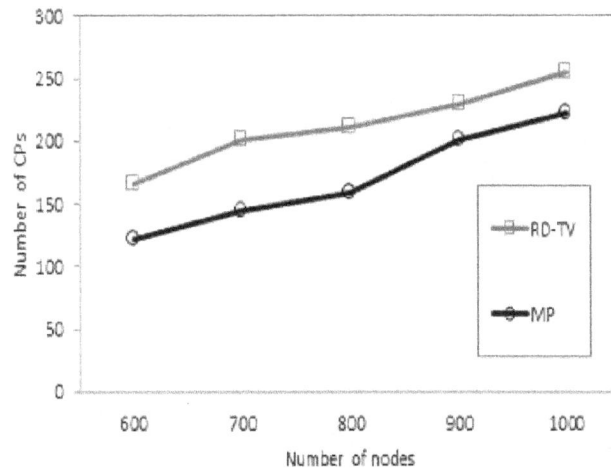

Figure 8: Number of caching points against the number of nodes, for the uniform density deployment scenario

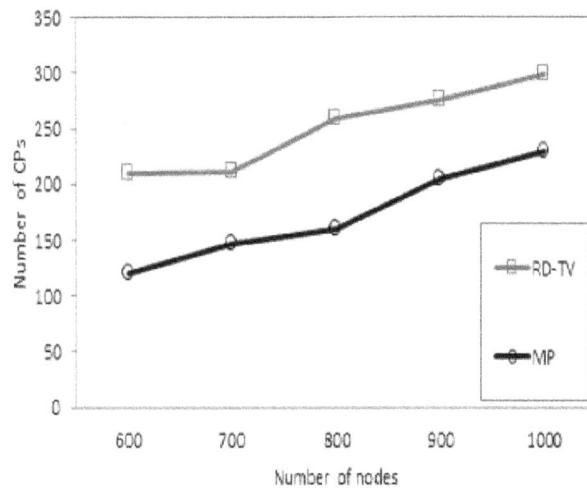

Figure 9: Number of caching points against the number of nodes, for the variable density deployment scenario

X. CONCLUSIONS

In this paper, we consider the problem of designing the mobile elements tours such that total size of the routing trees is minimized. In this work, we present an algorithmic solutions that creates its solution by partitioning the network, then in each partition; a caching node is selected based on the distribution of the nodes. An interesting open problem would be to consider application scenarios where the data gathering latency requirements vary in the network. For example, some areas in the network need to send data more frequently than others. In this case the tour length constraints would be different for different areas.

REFERENCES

[1]   J. Butler, "Robotics and Microelectronics: Mobile Robots as Gateways into Wireless Sensor Networks," Technology@Intel Magazine, 2003.

[2]   A. LaMarca, W. Brunette, D. Koizumi, M. Lease, S. Sigurdsson, K. Sikorski, D. Fox, and G. Borriello, "Making sensor networks practical with robots," Lecture notes in computer science, pp. 152–166, 2002. [Online]. Available: http://www.springerlink.com/index/9BUYG5K8BEW05CRX.pdf

[3]   E. Ekici, Y. Gu, and D. Bozdag, "Mobility-based communication in wireless sensor networks," IEEE Communications Magazine, vol. 44, no. 7, pp.56–62, 2006.

[4]   K. Dantu, M. Rahimi, H. Shah, S. Babel, A. Dhariwal, and G. Sukhatme, "Robomote: enabling mobility in sensor networks," in Proceedings of the 4th international symposium on Information processing in sensor networks (IPSN). IEEE Press, 2005, pp. 404–409. [Online]. Available: http://portal.acm.org/citation.cfm?id=1147685.1147751

[5]   R. Pon, M. Batalin, J. Gordon, A. Kansal, D. Liu, M. Rahimi, L. Shirachi, Y. Yu, M. Hansen, W. Kaiser, and Others, "Networked infomechanical systems: a mobile embedded networked sensor platform," in Proceedings of the 4th international symposium on Information processing in sensor networks. IEEE Press, 2005, pp. 376–381. [Online]. Available: http://portal.acm.org/citation.cfm?id=1147685.1147746

[6]   S. Gandham, M. Dawande, R. Prakash, and S. Venkatesan, "Energy efficient schemes for wireless sensor networks with multiple mobile base stations," in Proceedings of IEEE Globecom, vol. 1. IEEE, 2003, pp. 377–381.

[7]   Z. Wang, S. Basagni, E. Melachrinoudis, and C. Petrioli, "Exploiting sink mobility for maximizing sensor networks lifetime," in Proceedings ofthe 38th Annual Hawaii International Conference on System Sciences (HICSS), vol. 9, 2005, pp. 03–06. [Online]. Available: http://doi.ieeecomputersociety.org/10.1109/HICSS.2005.259

[8]   M. Zhao and Y. Yang, "Optimization-Based Distributed Algorithms for Mobile Data Gathering in Wireless Sensor Networks," IEEE Transactions on Mobile Computing, vol. 11, no. 10, pp. 1464–1477, 2012.

[9]   ——, "Efcient data gathering with mobile collectors and space-division multiple access technique in wireless sensor networks," IEEE Transactions on Computers, vol. 60, no. 3, pp. 400–417, 2011.

[10]   N. Christofides, "Worst-case analysis of a new heuristic for the traveling salesman problem," 1976.

[11]   Almahameed, M. A., M. Aalsalem, K. Almiani, and G. Al-Naymat, "Data Gathering with Tour Length-Constrained," Global Journal of Computer Science and Technology Network, vol. 13, pp. 41–51, 2013.

[12]   M. Aalsalem, K. Almi'ani, and B. Alqaralleh, in Proceedings of Computing, Communications and Applications Conference (ComComAp), 2012, pp. 278–282.

[13]   Y. Gu, D. Bozdag, E. Ekici, F. Ozguner, and C. Lee, "Partitioning based mobile element scheduling in wireless sensor networks," in Sensor and Ad Hoc Communications and Networks, 2005. IEEE SECON 2005. 2005 Second Annual IEEE Communications Society Conference on, 2005, pp. 386–395.

[14]   K. Almi'ani, M. Aalsalem, and R. Al-Hashemi, "Data gathering for periodic sensor applications," in Proceedings of the 2011 12th International Conference on Parallel and Distributed Computing, Applications and Technologies, ser. PDCAT '11, 2011, pp. 215–220.

[15]   K. Almi'ani, S. Selvadurai, and A. Viglas, "Periodic Mobile Multi-Gateway Scheduling," in Proceedings of the Ninth International Conference on Parallel and Distributed Computing, Applications and Technologies (PDCAT). IEEE, 2008, pp. 195–202. [Online]. Available: http://ieeexplore.ieee.org/lpdocs/epic03/wrapper.htm?arnumber=4710981

[16] M. Ma and Y. Yang, "Data gathering in wireless sensor networks with mobile collectors," in 2008 IEEE International Symposium on Parallel and Distributed Processing. IEEE, Apr. 2008, pp. 1–9. [Online]. Available: http://ieeexplore.ieee.org/lpdocs/epic03/wrapper.htm?arnumber=4536269

[17] K. Almi'ani, M. A. Abuhelaleh, and A. Viglas, "Length-constrained and connected tours for sensor networks," in Proceedings of the 2012 13th International Conference on Parallel and Distributed Computing, Applications and Technologies, ser. PDCAT '12. Washington, DC, USA: IEEE Computer Society, 2012, pp. 105–110. [Online]. Available: http://dx.doi.org/10.1109/PDCAT.2012.87

[18] M. Abuhelaleh, K. Almi'ani, and A. Viglas, "Connected tours for sensor networks using clustering techniques," in Proceedings of the 22nd Wireless and Optical Communication Conference (WOCC), 2013.

[19] K. A. Almi'ani, "Data Gathering with Tour-length Constrained Mobile Elements in Wireless Sensor Networks," Ph.D. dissertation, University of Sydney, Oct. 2010.

[20] A. Somasundara, A. Kansal, D. Jea, D. Estrin, and M. Srivastava, "Controllably mobile infrastructure for low energy embedded networks," IEEE Transactions on Mobile Computing, vol. 5, no. 8, pp. 958–973, 2006.

[21] R. Shah, S. Roy, S. Jain, and W. Brunette, "Data MULEs: Modeling a Three-tier Architecture for Sparse Sensor Networks," in Proceedings of the First IEEE Workshop on Sensor Network Protocols and Applications,   2003,   pp.   30–41.   [Online].   Available: http://www.comp.nus.edu.sg/~tayyc/6282/forPresentation/SensorNet.pdf

[22] D. Jea, A. Somasundara, and M. Srivastava, "Multiple controlled mobile elements (data mules) for data collection in sensor networks," Lecture Notes in Computer Science, vol. 3560, pp. 244–257, 2005. [Online]. Available: http://www.springerlink.com/index/4617cluarmakkx5x.pdf

[23] G. Xing, T. Wang, Z. Xie, and W. Jia, "Rendezvous planning in wireless sensor networks with mobile elements," IEEE Transactions on Mobile Computing, vol. 7, no. 12, pp. 1430–1443, Dec. 2008. [Online]. Available: http://ieeexplore.ieee.org/lpdocs/epic03/wrapper.htm?arnumber=4492781

[24] G. Xing, T. Wang, W. Jia, and M. Li, "Rendezvous design algorithms for wireless sensor networks with a mobile base station," in Proceedings of the 9th ACM international symposium on Mobile ad hoc networking and computing (MobiHoc). ACM New York, NY, USA, 2008, pp. 231–240. [Online]. Available: http://portal.acm.org/citation.cfm?id=1374650

[25] H. Salarian, C. Kwan-Wu, and F. Naghdy, "An Energy Efficient Mobile Sink Path Selection Strategy for Wireless Sensor Networks," IEEE Transactions on Vehicular Technology, vol. PP, 2013.

[26] K. Almi'ani, A. Viglas, and L. Libman, "Energy-Efficient Data Gathering with Tour Length-Constrained Mobile Elements in Wireless Sensor Networks," in Proceedings of the 36th IEEE Conference on Local Computer Networks (LCN). IEEE, 2010, pp. 598–605.

[27] K. Almi'ani, A. Viglas, and M. Aalsalem, "Mobile element path planning for gathering transit-time constrained data," in Proceedings of the 2011 12th International Conference on Parallel and Distributed Computing, Applications and Technologies, ser. PDCAT '11, 2011, pp. 221–226.

[28] K. Almi'ani and A. Viglas, "Designing connected tours that almost cover a network," in Proceedings of the 2013 14th International Conference on Parallel and Distributed Computing, Applications and Technologies, ser. PDCAT '13, 2013, p. in press.

[29] P. Toth and D. Vigo, "The Vehicle Routing Problem," Society for Industrial & Applied Mathematics (SIAM), 2001.

[30] J. Hou, L. Kung, N. Li, H. Zhang, W. Chen, H. Tyan, and H. Lim, "J-Sim: A Simulation and emulation environment for wireless sensor networks," IEEE Wireless Communications Magazine, vol. 13, no. 4, pp. 104–119, 2006.

[31] J. Hill and D. Culler, "Mica: A wireless platform for deeply embedded networks," IEEE micro, vol. 22, no. 6, pp. 12–24, 2002.

10

Performance Evaluation of AODV Routing Protocol in Cognitive Radio Ad-hoc Network

Prof.Shubhangi Mahamuni[1]
Dr.Vivekanand Mishra (Senior Member IEEE) [2]
Dr.Vijay M.Wadhai[3]

[1]Assist. Prof, Dept.of E&TC, MAE, Alandi (D), Pune,MS.
Shubhangimahamuni4@gmail.com
[2]Associate Professor, Dept.of Electronics, SVNIT, Surat,Gujrat.
vive2009@gmail.com
[3]Principal, MITCOE, Kothrud,Pune, MS
wadhai.vijay@gmail.com

Abstract—

A cognitive radio is designed for the utilisation of unused frequency. The transmission opportunity of a cognitive node is not guaranteed due to the presence of primary users (PUs).To better characterize the unique features of cognitive radio networks, we propose new routing metrics, including Routing for CRNs using IEEE 802.11 which are the official standards for wireless communication. Routing protocols, for network without infrastructures, have to be developed. These protocols determine how messages can be forwarded, from a source node to a destination node which is out of the range of the former, using other mobile nodes of the network. Routing, which includes for example maintenance and discovery of routes, is one of the very challenging areas in communication. Numerous simulations of routing protocols have been made using different simulators, such as ns-2.The impact of sensing time, route path and mobility in Ad- Hoc networks on connectivity and throughput tested.

Keywords: Ccognitive radio, CRAHNS (Cognitive Radio Adhoc Network), Spectrum sensing, Dynamic Spectrum Access, spectrum handoff.

I. Introduction

Cognitive Radio (CR) can be used as a solution to current unbalanced spectrum utilization. The cognitive ad hoc network can take advantage of dynamic spectrum access and spectrum diversity over wide spectrum. It could achieve higher network capacity compared to traditional ad hoc networks, thus supporting bandwidth-demanding applications. Mobile Ad- Hoc networking has gained an important part of the interest of researchers and become very popular these past few years, due to its potential and possibilities. These protocols determine how messages can be forwarded, from a source node to a destination node which is out of the range of the former, using other mobile nodes of the network. Routing, which includes for example maintenance and discovery of routes, is one of the very challenging areas in communication. Simulators though cannot take into account of all the factors that can come up in real life and performance and connectivity of mobile Ad-Hoc network depend and are limited also by such factors. Here we have tried routing protocol and end to end protocol to enhance throughput of a cognitive radio adhoc network .The impact of varying packet size, route length and mobility in Ad- Hoc networks on connectivity, Route Discovery Time and throughput is tested. In this paper we included an introduction to Ad-Hoc Mobile Networks and IEEE 802.11in part I. Part II consists routing in CRAHNs. Simulation flowchart, algorithm, results and software are described in chapter III.

A. Cognitive Radio Ad-Hoc Network

Cognitive Radio Ad-Hoc Network (CRAHNs)[1, 2] is a new developed technology of wireless communication. The difference to traditional wireless networks is that there is no need for established infrastructure. Since there is no such infrastructure and therefore no preinstalled routers which can, for example, forward packets from one host to another, this task has to be taken over by participants, also called mobile nodes, of the network. Each of those nodes takes equal roles, what means that all of them can operate as a host and as a router. Traditional wireless networks are need some improvement due to some factors such as security, power control, transmission quality and bandwidth optimization. To solve problems like maintenance and discovery of routes and topological changes of the network is the challenge of Ad-Hoc Networking.

We have formed cognitive radio adhoc network by using user nodes and communication channels as shown in fig.1.

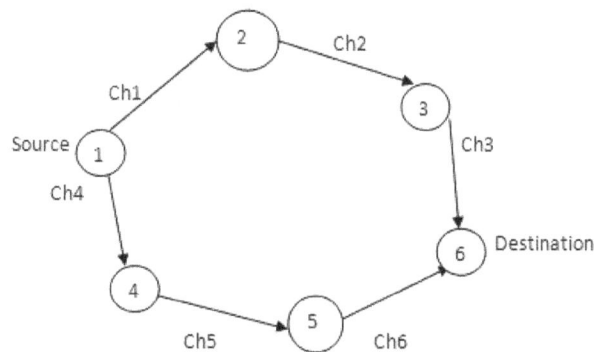

Fig.1 Cognitive Radio Ad-Hoc Network

B. RELATED WORK

A number of routing protocols have been proposed and implemented for CRAHNs in order to enhance the bandwidth utilization, higher throughputs, lesser overheads per packet, minimum consumption of energy and others. All these protocols have their own advantages and disadvantages under certain circumstances. The major requirements of a routing protocol was proposed by Tao Lin et al.[2] that includes minimum route acquisition delay, quick routing reconfiguration, loop-free routing, distributed routing approach. A review of Routing Protocols has been explained in detail by Changling Liu, Jorg Kaiser[3,4,5,6].Obviously, most of the routing protocols are qualitatively enabled. A lot of simulation studies were carried out in the paper [2] to review the quantitative properties of routing protocols. The idea of the Cognitive radio Adhoc network was explained by I. Akyildiz, W.Y. Lee, K.R. Chowdhury[1]. A number of extensive simulation studies on various CRAHNs routing protocols have been performed in terms of control, route discovery and route maintenance[7,8][9,10]. However, there is a severe lacking in implementation and operational experiences with existing CRAHNs routing protocols. The various types of mobility models were identified and evaluated by K.R. Choudhury, M.D. Felice et al. [17] because the mobility of a node will also affect the overall performance of the routing protocols. A framework for the ad hoc routing protocols was proposed by Tao Lin et al. [3] using Relay Node Set which would be helpful for comparing the various routing protocols like AODV, OLSR & TBRPF [17]. Application of the standards like IEEE 802.11,TCP has explained in detail by M. Gandetto, C. Regazzoni and Goff, J. Moronski et al. [13,14]. The performance of the routing protocols AODV by considering the metrics of packet delivery ratio, control traffic overhead and route length by using NS2 simulator [18][19][20][21].

II.AODV Protocol and Routing

A.AODV Protocol:

AODV protocol is a routing protocol in a reactive routing protocol. Routing is only when needed. Fig.1shows message transmission using AODV protocol. Routing in a Cognitive Radio Adhoc Network (CRAHNs) is done with the goal of finding a short and optimized route from the source to the destination node. An advantage is that smaller bandwidth is needed for maintaining routing tables, and disadvantage is it will create non negligible delay, since before using the route for a specific communication, it has to be determined. When a source has to transmit to an unknown destination sends RREQ route request for the destination [3,4]. If the receiving node has not received RREQ,it is not the destination node and does not have the current route to the destination.RREP of used for hopping purpose from the souse towards destination, it generates route reply(RREP).for the creation of routes between the nodes RREP is used. As shown in fig.2 S is the source and D is the destination used for the transmission of simple message Hello towards destination D.

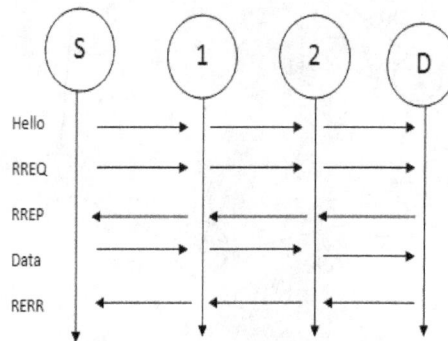

Fig.2 AODV Protocol Messaging

When a route is needed and it is not already known by a node it sends a Route Request (RREQ) message to its neighbours. Those forward the message until it reaches the destination node. Each intermediate node updates the RREQ with its address. When a node sends a RREQ message it attaches to it a request ID. The request ID and the IP address of the node form a unique identifier. This is done in order to prevent that an intermediate node which receives twice the same RREQ message forwards it twice. When the destination receives the RREQ it sends a Route Reply (RREP) message back to the source node by reversing the hop sequence recorded in the RREQ. When the source receives the RREP it can start communicating with the destination, by including the whole route in the header of each to be sent message.

B. IEEE 802.11

The scope of IEEE 802.11 [5] is to develop and maintain specifications for wireless connectivity for fixed, portable and moving stations within a local area. It defines over-the-air protocols necessary to support networking in a local area. This standard provides MAC and physical layer functionality.

IEEE 802.11 can be used in two different operating modes:

1. Infrastructure mode
2. Independent (Ad-Hoc) mode

At the MAC layer 802.11 uses both carrier sensing and virtual carrier sensing before sending data to avoid collision. Use of RTS and CTS is used for virtual carrier sensing. The network allocation vector (NAV) is used to perform virtual channel sensing for the indication that the channel is busy.

ACK is used for the indication that the destination received data correctly. This way RTS-CTS-Data-ACK is called Distributed Coordination Function (DCF).

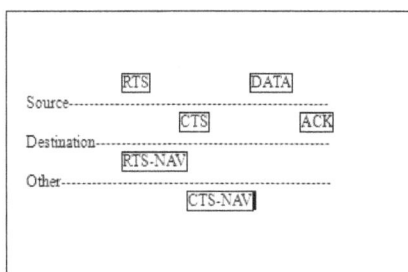

Fig.3 IEEE 802.11Disributed co-ordination Function

III.Routing In a CRAHNs

A. Route Maintenance:

When a route has been established, it is being maintained by the source node as long as the route is needed. Movements of nodes affect only the routes passing through this specific node and thus do not have global effects. If the source node moves while having an active session, and loses connectivity with the next hop of the route, it can rebroadcast an RREQ. If though an intermediate station loses connectivity with its next hop it initiates an Route Error (RERR) message and broadcasts it to its precursor nodes and marks the entry of the destination in the route table as invalid, by setting the distance to infinity[6,7]. The entry will only be discarded after a certain amount of time, since routing information may still be used for the transmission of message from source.Fig.3 shows the shifting of path for transmission of data from different source and destination.

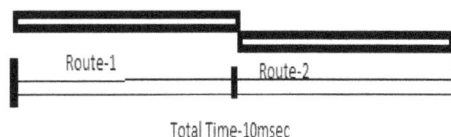

Total Time-10msec
Fig.4.Channel Utilisation in CRAHNs

B.Transport layer:

Since TCP is the de facto transport protocol standard on Internet, it is crucial to estimate its ability in providing stable end-to-end communication over CRAHNs.In this set of experiments, the CRAHN environment is based on four channels (i.e. $N = 4$) and constructed as follows We study the performance of TCP under different CRAHNs characteristics, e.g. the sensing time interval of CR users. The choice of the single-hop scenario can be motivated as follows. First, the single-hop scenario is simple enough to understand the impact of CRAHNs characteristics on the dynamics of TCP, while this might be difficult to investigate in multi-hop topologies. Second, the single-hop topology constitutes a base case, from the point of view of protocol performance. If we discover that a single parameter, e.g. the sensing time interval, has a strong impact on TCP performance, then this effect will be emphasized in a multi-hop environment by the presence of multiple intermediate nodes between the source and the destination CR users. Moreover, although very simple, the single-hop topology constitutes a realistic model for the evaluation of infrastructure-based CR networks, where the mobile CR users are attached to a fixed cognitive base station (BS).

We consider two metrics in the performance analysis:

• TCP throughput: This is the end-to-end TCP throughput at the application layer, i.e. the amount of bits for seconds received by the upper layer FTP application at the destination node, without considering out-of-order, duplicated and TCP–ACK packets.

• TCP efficiency: This is an estimation of bandwidth resource utilization by TCP. It is defined as follows:

$$\epsilon = \frac{TCP_{THR}(t_1, t_2)}{\int_{t_1}^{t_2} C(t) \cdot dt}, \quad 0 \leq \epsilon \leq 1$$

(5)

where $TCP_{THR}(t1, t2)$ is the average TCP throughput computed over the measurement period from t, and $C(t)$ is the available channel capacity at time t .All the metrics described so far are measured at CR users since the focus here is to evaluate the impact of PU activity on the performance of the TCP protocol for secondary systems.

C.Sensing Cycle Analysis Simulation Setup:

In the sensing cycle analysis, we consider the basic single-hop topology, and we vary the sensing interval (*ts*) and operation interval of CR users. All the primary bands have capacity equal to 5 Mb/s.

IIV. SIMULATION AND RESULTS:

A. Flowchart:

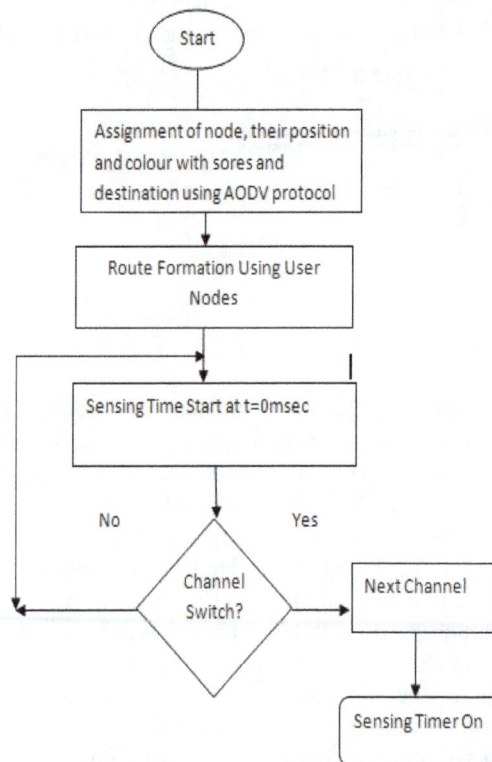

Fig.5. Flowchart for channel allocation and routing in Cognitive Radio Ah-doc Networks

B. Node Mobility:

In CR ad-hoc networks, TCP-EFLN and A TCP react to the route disruption after it happens by an explicit notification in the form of the internet control message protocol (ICMP) message at the IP layer[8,9]. For CRAHNs, intermediate nodes may continue their periodic sensing if a route failure is detected at a further downstream node. In such cases, the route failure message is delayed at each hop that undertakes sensing and the source is informed much later. [10] and hence a predictive mobility model needs to be incorporated in the TCP rate control mechanism. Two broad approaches may be adopted in the design of transport layer protocols [11, 12].

(i)   The standard TCP and UDP protocols may be adapted by making them channel aware and sensitive to PU activity.

(ii)   Scenario and application specific protocols may be devised those tradeoffs the generality in implementation for optimum performance under known channel condition.

C. Implementation of AODV Protocol

Performance evolution of the AODV protocol is implemented on the Linux 2.4 Kernel with the following considerations.

Parameter	Value
Channel Type	Channel/Wireless Channel
propagation	Propagation/Two Ray Ground
antenna Type	Antenna/Omni Antenna
Network Layer	LL
Queue	Queue/Drop Tail/PriQueue
Queue Length	200
Network Layer	Phy/WirelessPhy
MAC protocol	Mac/802_11
No. of Nodes	27
Radio Protocol	AODV
Value (x)	500
Value (y)	500
ns	new Simulator

Table 1. Configure and Create Nodes N=7

3. Define color index
4. Define source and destination positions
5. Set up a flow: Flow of a route can be set up by using following channels.
set chan_1 [new $val(chan)]
set chan_2 [new $val(chan)]
set chan_3 [new $val (chan)]
set chan_4 [new $val (chan)]
6. Request data
7. Execute simulation.

IV.Simulation Results:

Following are the real-time snapshots taken for the calculation of throughput .Fig.6, Fig.7 and Fig.8shows transmission of message at different time periods. The green color nodes of the fig.6 indicates one route for the transmission of packets from source to destination..Simultaneously sensing of the nodes along with the transmission of packet takes place.Fig.7 shows state of changing route. Whereas fig.8 shows blue color nodes frorn different route along with the sensing of the nodes. Throughput is the ratio of the total amount of data that reaches a receiver from a sender to the time it takes for the receiver to get the last packet [13, 14, 15, and 16]. It measures of effectiveness of a routing protocol. The throughput values of AODV Protocols for 7 Nodes at Pause time 10s are noted and they are plotted on the different scales to best show the effects of varying throughput of the above routing protocols (Figures 6, 7 & 8). Based on the simulation results, the throughput value of AODV increases initially and reduces when the time increases [17, 18, 19]. The throughput value of AODV slowly increases initially and maintains its value when the time increases.

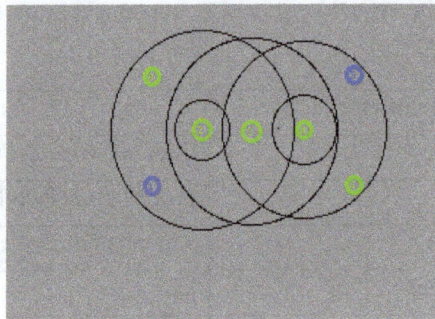

Fig.6. Initially with N0 as a source and N5 destination

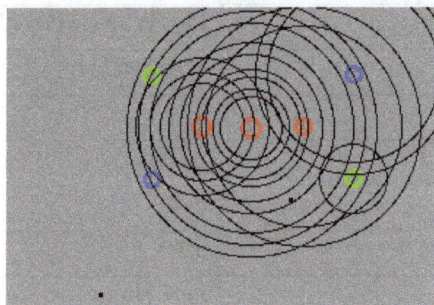

Fig.7. Packet transmission with spectrum sensing

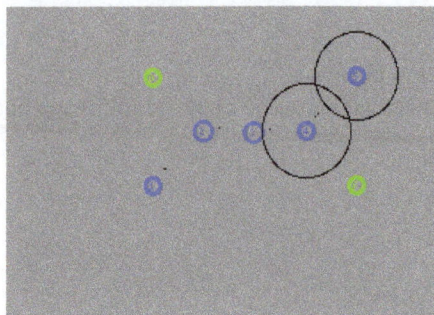

Fig.8. Initially with N1 as a source and N6 destination

The duration of the periodic spectrum sensing decides, in part, the end-to-end performance a shorter sensing time may result in higher throughput but may affect the transport layer severely if a PU is misdirected.

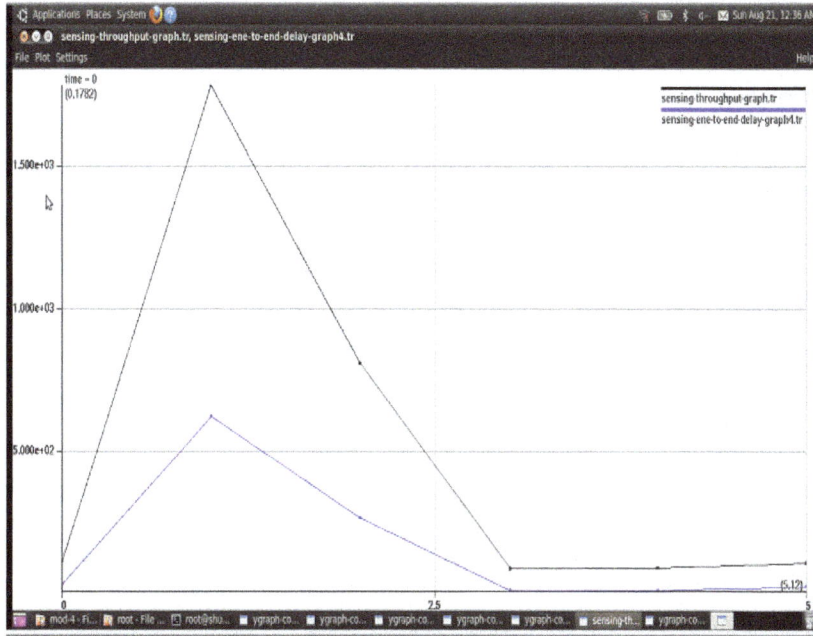

Fig.9.Throughput for CRAHNs with respect to sensing Time with End to End Delay

Conclusion:

Thus, the sensing scheme needs to be integrated in the design of the transport protocol. Fig.9 shows throughput of a CRAHNs.It clearly shows that the throughput with routing is more than the throughput without routing. This analysis of throughput will be required whenever there is a requirement of the delay calculation in case of handoff mechanism. As during the handoff mechanism during routing process switching delay and node delay occurs and that will be the challenge for the researchers.

REFERENCES:

[1]   I. Akyildiz, W.Y. Lee, K.R. Chowdhury, CRAHNs: Cognitive Radio Ad HoC Networks, Ad Hoc Networks Journal 7,2009, 810–836.

[2]   Tao Lin, Scott F.Midkiff and Jahng S.Park ,A framework for Wireless Ad hoc Routing Protcols, IEEE 2003.

[3]   Zuraida Binti Abdullah Hani and Mohd. Dani Bin Baba, Designing Routing protocols for Mobile Ad hoc networks,IEEE 2003.

[4]   Dhiraj Nitnaware, Ajay Verma, Energy constraint Node cache based routing protocol for Adhoc Network, IJWMN, Feb. 2010.

[5]   Mehran Abolhasan, Tadeusz Wysoci, Eryk Dutkiewicz, A review of routing protocols for mobile ad hoc networks, ELSEVIER , 2003.

[6]   Changling Liu, Jorg Kaiser, A survey of Mobile Ad Hoc Network Routing Protocols, University of Magdeburg, 2005.

[7]   Yih-Chun Hu and David B.Johnson ,Caching Strategies in On-Demand Routing Protocols for Wireless Ad hoc Networks, ACM 2000.

[8]   Mehran Abolhasan, Tadeusz Wysocki and Eryk Dutkiewicz , A review of routing protocols for mobile ad hoc networks, Elsevier 2003.

[9]   K.R. Chowdhury, I.F. Akyildiz, Cognitive wireless mesh networks with dynamic spectrum access, IEEE Journal of Selected Areas in Communications 26, 2008,168–181.

[10] K.R. Chowdhury, M. Di Felice, Search: a routing protocol for mobile cognitive radio ad-hoc networks, Computer Communication Journal 32, 2009, 1983-1997.

[11] K. Akkaya, M. Younis, A survey on routing protocols for wireless sensor networks, Ad Hoc Networks (Elsevier) 3,2005, 325–349

[12] C.Perkins, Ad hoc on-demand distance vector (AODV) routing, RFC 3561,July 2003

[13] M. Gandetto, C. Regazzoni, Spectrum sensing: a distributed approach for cognitive terminals, IEEE Journal on Selected Areas in Communications 25,2007, 546–557.

[14] Goff, J. Moronski, D.S. Phatak, V. Gupta, Freeze-TCP: a true end-to-end TCP enhancement mechanism for mobile environments, in: Proceedings of the IEEE INFOCOM, Tel-Aviv, Israel, 2000,1537–1545.

[15] L. Hanzo (II.), R. Tafazolli, A survey of QoS routing solutions for mobile ad hoc networks, in: IEEE Communications Surveys and Tutorials, vol. 9 (2), 2007,50–70.

International Journal of Wireless & Mobile Networks (IJWMN) Vol. 3, No. 5, October 2011

[16] K. Chandran, S. Raghunathan, S. Venkatesan, R. Prakash, A feedback based scheme for improving TCP performance in ad hoc wireless Networks, IEEE Personal Communications Magazine 8,2001, 34– 39.

[17] K.R. Chowdhury, M.D. Felice, SEARCH: a routing protocol for mobile cognitive radio ad-hoc networks,2008,313-320.

[18] C. Cordeiro, K. Challapali, C-MAC: a cognitive MAC protocol for multi-channel wireless networks, in: Proceedings of the IEEE DySPAN,2007, 147–157.

[19] H.M. El-Sayed,O.Bazon and U.Qureshi and M.Jaseemuddin Performance Evaluation of TCP in Mobile Ad Hoc Networks.

[20] Network simulator version 2.34 http://www.isi.edu/nsnam/ns/.

L. Cao, H. Zheng, Distributed spectrum allocation via local bargaining, in: Proceedings of the IEEE Sensor and Ad Hoc Communications and Networks (SECON), 2005, 475–486.

[21] Saiful Azadm, Arafatur Rahman and Farhat Anwar, A Performance comparison of Proactive and Reactive Routing protocols of Mobile Ad hoc Networks(MANET)), Journal of Engineering and Applied Sciences, 2007.

11

EFFECTS OF FILTERS ON THE PERFORMANCE OF DVB-T RECEIVER

AKM Arifuzzaman[1], Rumana Islam[1], Mohammed Tarique[2], and
Mussab Saleh Hassan[2]

[1]Department of Electrical and Electronic Engineering
American International University-Bangladesh, Banani, Dhaka, Bangladesh
E-mail:arifuzzaman,rumana@aiub.edu

[2]Department of Electrical Engineering
Ajman University of Science and Technology, Fujairah, UAE
E-mail:m.tarique,m.mohammad@ajman.ac.ae

ABSTRACT

Digital Video Broadcasting-Terrestrial (DVB-T) is an international standard for digital television services. Orthogonal Frequency Division Multiplexing (OFDM) is the core of this technology. OFDM based system like DVB-T can handle multipath fading and hence it can minimize Inter Symbol Interference (ISI). DVB-T has some limitations too namely large dynamic range of the signals and sensitivity to frequency error. In order to overcome these limitations DVB-T receivers should be optimally designed. In this paper we address the issues related to optimal DVB-T receiver design. There of several signal processing units in a DVB-T receiver. A low-pass filter is one of them. In this paper, we consider some classic filters namely Butterworth, Chebyshev, and elliptic in the DVB-T receiver. The effects of different filters on the performances of DVB-T receiver have been investigated and compared in this paper under AWGN channel condition.

KEYWORDS

DVB, DVB-T, multi-carrier, orthogonal, FFT, IFFT, BER, ISI, AWGN, Butterworth, Elliptic, Chebyschev, PSD

1. INTRODUCTION

Television (TV) is considered as the most cost-effective source of entertainment, education, and information [1-2]. The International Telecommunication Union (ITU) has estimated 1,416,338,245 TV sets in the households around the World [3]. Originally TV was introduced as an electromechanical system [2,4]. An all-electronic system was developed in Europe in the twentieth century. During this time period the Federal Communication Commission (FCC) adopted the recommendations made by the National Television System Committee (NTSC) system in USA [3]. Later on a number of color television systems were introduced. The Sequential Couleur A Memoire (SECAM) and Phase Alternating Line (PAL) were introduced in Europe. NTSC color television was introduced in USA. The digital television network merged with the advancements of digital technology. Broadcasters and manufacturer realized the importance of switching from analog to digital television. Digital television became popular because of its improved spectrum efficiency. The television broadcasters could accommodate more channels in a limited spectrum. Several digital TV standards were rapidly developed in the next few years. In Europe broadcasters and electronic companies formed consortium and they introduced Digital Video Broadcasting (DVB). It is a set of internationally recognized standards for digital television services. It is maintained by DVB project [5]. This project is an industry-led consortium of more than 200 manufacturers, broadcasters, network operators, software developers, and regulatory bodies. European Broadcasting Union, European Committee for

Electrotechnical   Standardization,   and   Joint   Technical   Committee   of   European Telecommunication Standards Institute publish and update this standard. This standard has been proposed to support digital television services to the consumers by means of rooftop antennas. There are three key DVB standards that define the physical and data link layer of the distribution system. These are (i) DVB-S for satellite network, (ii) DVB-C for cable networks, and (iii) DVB-T for terrestrial networks [6-8]. There are also some related standards. Some of these are DVB-H and DVB-IPTV. The DVB-H has been standardized to support mobile television services in handheld devices. On the other hand, DVB-IPTV has been standardized to support television service by using Internet Protocol (IP) [9]. Among all these standards DVB-T system has become the most popular one and is the topic of this investigation. DVB-T system has been formally defined and introduced in [10]. The system is designed to operate within the existing Very High Frequency Band (50-230 MHz) and Ultra High Frequency Band (470-870 MHz). Two modes of operation have been defined for DVB-T standard namely "2K mode" and "8K mode". The "2K mode" has been defined for DVB-T transmission and the "8K mode" has been defined for DVB-H transmission [11]. For single transmission and limited distance operation the "2K mode" is considered suitable. On the other hand, the "8K mode" is preferable for long distance operation. The DVB-T standard allows different levels of Quadrature Amplitude Modulation (QAM) and it uses different inner code rates. The system also allows two levels of hierarchical channel coding and modulation. The basic functional block diagram of DVB standard is shown in Figure 1. The processes applied to the input data stream are (i) transport multiplex adaptation, (ii) outer coding, (iii) outer interleaving, (iv) inner coding, (v) inner interleaving, (vi) mapping and modulation, and (vii) OFDM transmission.

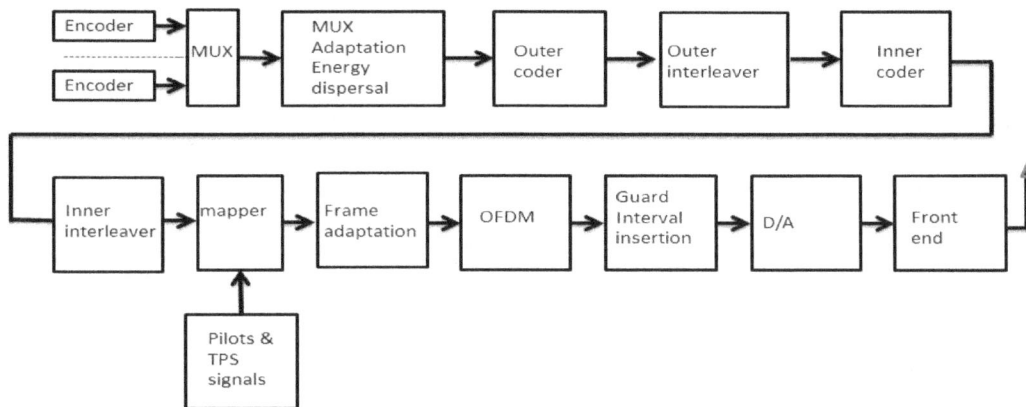

Figure 1. Block ddiagram of DVB system.

Most of the signal processing steps shown in Figure 1 is carried out in a digital signal processor. The performance of DVB system highly depends on the modulation. Since this investigation addresses the physical layer issues the basic components we investigate here are the OFDM, Digital-to-Analog Converter (D/A) and the front end. Brief descriptions of these components are provided here for the completeness of this paper.

2. OFDM MODULATION

The Orthogonal Frequency Division Multiplexing (OFDM) is a multi-carrier modulation (MCM) scheme [12-13]. The main principle of MCM is to divide the high rate input bit stream into several low rate parallel bit streams and then these bit streams are used to modulate several sub-carriers. The sub-carriers are separated by a guard band to ensure that they do not overlap with each other during the course of transmission. In the receiver side band-pass filters are used to separate the spectrum of individual sub-carriers. OFDM is a special form of spectrally efficient MCM technique, which employs densely spaced orthogonal sub-carriers and overlapping spectrums. Hence, it is considered as a spectrum efficient modulation scheme.

Unlike other systems OFDM does not require band-pass filters because of the orthogonality nature of the subcarriers. Hence the available bandwidth is used very efficiently without causing the Inter Carrier Interference (ICI). The required bandwidth can be further reduced by removing the guard band and allowing the subcarriers to overlap. Despite the overlapping spectrum it is still possible to recover the sub-carriers provided that the orthogonality condition among the subcarriers is maintained. OFDM provides a composite high data rate with long symbol duration. Depending on the channel coherence bandwidth OFDM reduces or completely eliminates the effect of Inter Symbol Interference (ISI). The Discrete Fourier Transform (DFT) and Inverse Discrete Fourier Transform (IDFT) can be used to implement the orthogonal signals. OFDM has numerous advantages over other digital modulation schemes. The main advantage of OFDM is that it can minimize the Inter Symbol Interference (ISI), which is originated from the multipath signal propagation [14] as mentioned earlier. Hence, it can support a very high data rate. There are some other advantages too. OFDM can cope with the multipath fading with less computational complexity compared to other modulation schemes. It is more resistant to frequency selective flat fading. It uses computationally efficient Fast Fourier Transform (FFT) and Inverse Fast Fourier Transform (IFFT) algorithms. OFDM is also considered suitable for coherent demodulation. Additionally, OFDM can ensure the required Quality of Service (QoS) to the end users depending on the bandwidth. It is also considered suitable for diversity techniques (i.e., time diversity and frequency diversity). The orthogonality of subcarriers of an OFDM system can be jeopardized when the same is sent through the multipath channel, which also introduces Inter Carrier Interference (ICI). Cyclic Prefix (CP) is added with the OFDM symbol to combat both ISI and ICI. The CP contains a copy of the last part of the OFDM symbol appended to the front of transmitted OFDM symbol. The length of the CP must be longer than the maximum delay spread of the multipath environment to reduce the ISI. OFDM has some disadvantages too. Some of these are: (i) it requires a strong synchronization between the transmitter and the receiver, (ii) it is very much sensitive to phase noise and frequency offset, (iii) it is not power efficient because of the always active Fast Fourier Transform (FFT) algorithm and Forward Error Correction (FEC), (iv) it cannot take the advantages of the diversity gain if few sub-carriers are allotted to each user, and (v) it requires a high peak to average ration RF power amplifiers to avoid amplitude noise. There have been numerous works carried out to overcome these disadvantages. We have included some of them here that are related to this work.

3. RELATED WORKS

The bandwidth efficiency of OFDM systems is originated from the overlapping orthogonal sub-carriers. Traditionally it is assumed that the waveforms of OFDM transmission have band limited and the spectrums overlap only with its adjacent sub-carrier. But, in real life scenario the OFDM transmission is not band limited. Hence, the spectrums not only interfere with the adjacent carrier, but they also interfere with the further-away sub-carriers [15]. To limit this type of interference the spectrum of OFDM transmission must be kept in a sufficiently low stop-band. Many research works and innovations have been done to minimize the interference. One of the early solutions has been presented in [16]. The authors suggested some analysis which led to use truncated Prolate Spheroidal Wave Function (PSWF) [17]. Non-linear programming technique has been used to design FIR filters for transmitter and receiver in [18]. Pulse shaping and filtering have been suggested to limit the bandwidth of OFDM symbol. Raised Cosine Function (RCF) has been used in [19-22] to reduce the inter symbol interference (ISI) in OFDM system. The authors have shown in these works that BER can be reduced by using RCF filters. It is shown in [23] that reconstruction filters, anti-aliasing filters and other filters cause smearing in an OFDM system. The author presented some numerical works in the same work and they suggested that Chebyshev II filter approximations should be the best choice. This type of filter causes the least smearing. A series of investigations also shows that filtering is important to reduce Peak-to-Average Ration (PAR) of OFDM system. The authors suggested in [24] that the clipping is the simplest method to reduce PAR reduction. But, it causes both in-band distortion and out of band distortion. Hence, a filter should be used after the clipping to reduce the distortions. A more efficient method for clipping and filtering has been proposed in [25,26]. In these works the authors demonstrated that the clipping and filtering algorithm are better than clipping along to reduce PAPR in OFDM system. An easy to implement method of clipping and filtering has been discussed in [27]. The effects of performing channel estimation of an OFDM system by using non-ideal interpolating and decimating filters have been investigated under both AWGN and Rayleigh fading channel in [28]. Two schemes for ICI reduction have been proposed in [29]. In self-cancellation scheme redundant data is transmitted onto adjacent subcarriers to reduce ICI. In the second scheme an Extended Kalman Filter (EKF) corrects the frequency offset at the receiver. In a related work [30] an unscented Kalman filter based solution has also been provided. In contrast to other related works the authors considered high mobility OFDM system in this work. The effects of imperfect anti-aliasing filtering have been investigated in this paper [31]. The authors have used a linear phase FIR filter of order N=50 in this investigation. In [32] it has been claimed that pulse shaping filter is not good enough for ICI cancellation in OFDM system. A hybrid scheme consisting of pulse shaping and Maximum Likelihood Estimation (MLE) should be combined to cancel ICI. In [33] a non-linear adaptive filter has been proposed. The authors claimed that non-linear filter scheme is superior to the conventional linear filter.

In all the above mentioned works the investigators focused on a particular filter to improve the performance of OFDM system. Only a few works provided a comparative performance analysis of different types of filter. In these work we considered some classic filters namely Butterworth filter, Elliptic filters, and Chebyshev filters. We varied some filter parameters like cut-off frequency and filter order to investigate the performance of DVB-T receiver. It is worthwhile to mention another difference here. In all the above mentioned works basic OFDM system has been used. In this paper we consider DVB-T system, which uses OFDM as a modulation scheme.

4. DVB-T FRAME STRUCTURE

The DVB-T transmitted signals are organized into frame structure. Each frame consists of 68 OFDM symbols [9]. Since OFDM is the core of DVB-T system, we will use DVB-T symbol and OFDM symbol interchangeably in the remaining portion of this paper. Each OFDM symbol consists of a set of K=1705 carriers in the "2K mode" and K=6817 carriers for "8K mode". Let us denote the frame duration by T_F and the symbol duration by T_S. The symbol duration has two components namely a useful part and a guard interval. Let us denote the useful component by T_U and the guard component by T_G. The guard interval carries cyclic prefix (CP) inserted before the useful component (as mentioned earlier). Four guard intervals have been chosen in DVB-T standard namely 1/4, 1/8, 1/16 and 1/32. The symbols of OFDM are numbered from 0 to 67. In DVB-T standard all symbols contain data and reference information. The OFDM symbol can be considered as a combination of many separately modulated carriers. Hence each symbol can be considered to be divided into cells. The transmitted data of an OFDM frame contains: (i) scattered pilot cells, (ii) continual pilot carriers, and (iii) Transmission Parameter Signaling (TPS) carriers. The pilots are used for frame synchronization, frequency synchronization, time synchronization, channel estimation, and transmission mode identification. The TPS carriers are used for the purpose of signaling parameters (i.e., channel coding and modulation) related to the transmission scheme, The carriers are indexed from K_{min} to K_{max}, where K_{min}=0 and K_{max}=1704 for "2K mode" and K_{max}=6816 in "8K mode". The spacing between K_{min} and K_{max} is defined by $(K-1)/T_U$, where the spacing between adjacent carriers is $1/T_U$ and K is the number of the carriers. The numerical value for the OFDM parameters for the "2K mode" and "8K mode" are listed in Table 1.

Table 1. OFDM Parameters for "2K" and "8K" mode

Parameter	8K mode	2K mode
Number of carriers K	6817	1705
Value of carrier number K_{min}	0	0
Value of carrier number K_{max}	6816	1704
Duration T_U	896 µs	224 µs
Carrier spacing $1/T_U$	1116 Hz	4464 Hz
Spacing between K_{min} and K_{max} $(K-1)/T_U$	761 MHz	761 MHz

The transmitted OFDM symbol can be mathematically expressed by

$$s(t) = \mathrm{Re}\left\{ e^{2\pi f_c t} \sum_{m=0}^{\infty} \sum_{l=0}^{67} \sum_{k=K_{min}}^{K_{max}} c_{m,l,k} \times \psi_{m,l,k}(t) \right\} \tag{1}$$

where

$$\psi_{m,l,k}(t) = \begin{cases} e^{j2\pi \frac{k'}{T_U}(t-T_G-l\times T_S-68\times m\times T_S)} & (l+68\times m)\times T_S \le t \le (l+68\times m+1)\times \end{cases}$$

$$\psi_{m,l,k}(t)=0 \quad else$$

where k = the carrier number, l = the OFDM symbol number, m = the frame number, K = the number of transmitted carrier, f_c = the central frequency, $k\square$ = the carrier index related to the canter frequency, and $c_{m,l,k}$ = complex symbol for carrier k of the data symbol number l in frame m. The $c_{m,l,k}$ values are normalized according to the constellation points of the modulation alphabet used. The other parameters of DBV system are listed in Table 2.

Table 2. Additional Parameters

Mode	8K mode				2K mode			
Guard interval T_G/T_U	1/4	1/8	1/16	1/32	1/4	1/8	1/16	1/32
Duration of symbol part T_U				8192xT =896 µs				2048xT =224 µs
Duration of Guard Interval T_G	2048xT =224 µs	1024xT =112 µs	512xT =56 µs	256xT =28 µs	512xT =56 µs	256xT =28 µs	128xT =14 µs	64xT =7µs
Symbol Duration T_S	10240 xT 1120 µs	9216xT 1008 µs	8704xT 952 µs	8448xT 924 µs	2560xT 280 µs	2304xT 252 µs	2176xT 238µs	2112xT 231µs

5. DVB TRANSMITTER AND RECEIVER MODEL

The transmitter and the receiver models proposed in [34] have been used in this investigation. The basic signal processing steps done in the transmitter and the receiver of DVB-T system are shown in Figure 2. Since the number of carriers in DVB-T system is 1705, we considered 1705 4-QAM symbols as the input data to the transmitter. We used a bandwidth of 8.0 MHz. The carrier frequency we chose is 90 MHz. We divided the 8.0 MHz into 1705 sub-carriers. Since the OFDM symbol duration is specified considering a 2048-point Inverse Fast Fourier Transform (IFFT), we chose 4096-point IFFT. Hence, 2391 zeros were padded with the information symbol. The output carriers of IFFT were then converted into continuous time signal in the next signal processing step as shown in Figure 2. This step of the signal processing is accomplished by using a pulse shaping filter and a low pass filter [35] denoted by Digital to Analog (D/A) filter in the block diagram shown in Figure 2(a). The pulse shaping filter converts the discrete time carrier output signals into continuous signal. In the next step a D/A filter with a sharp bandwidth is used. We choose a Butterworth filter for this purpose. Finally, the carrier modulation is performed at the last stage.

Designing an OFDM based DVB receiver has been an open research issue since its conception. Like this work most of the researches conducted and innovations have been done in the DVB receiver design. The DVB receiver model used in this investigation is shown in Figure 2(b).

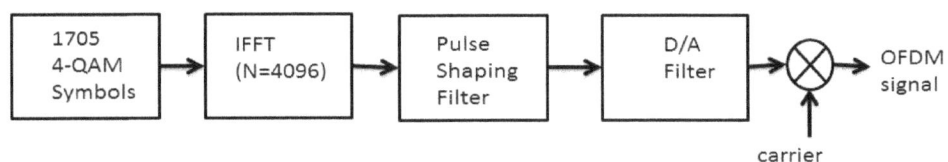

(a) Transmitter model

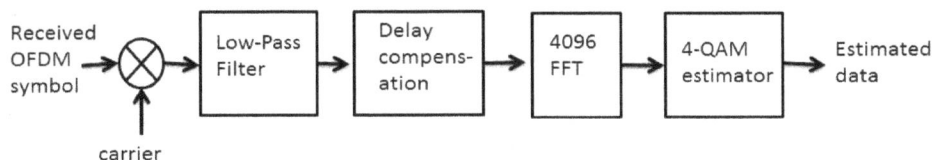

(b) Receiver model

Figure 2. OFDM transmitter and receiver model

The DVB receiver operation performs the inverse operation of the transmitter. First, the coherent demodulation is done. The demodulated OFDM signal is then passed through the low-pass filter to recover the continuous version of the OFDM signal. The delays produced by the reconstruction and demodulation filters are compensated. The resultant signal is then sampled to convert continuous OFDM symbol into its discrete version. The resulted signal is then passed through a 4096-FFT to do the inverse function of its counterpart (i.e., IFFT) done in the transmitter. Finally, the resultant data is estimated by using a 4-QAM estimator. The power spectral density (PSD) of DVB symbol is shown in Figure 3. A numerous investigations show that a significant level of OFDM spectrum falls outside the nominal bandwidth as shown in Figure 3. Hence an appropriate filter should be carefully selected in the transmitter and the receiver to minimize this out of band spectrum. In this investigation three classic filters namely Butterworth filter, Chebyshev filter, and elliptic filters have been considered and the DVB receiver performances have been investigated.

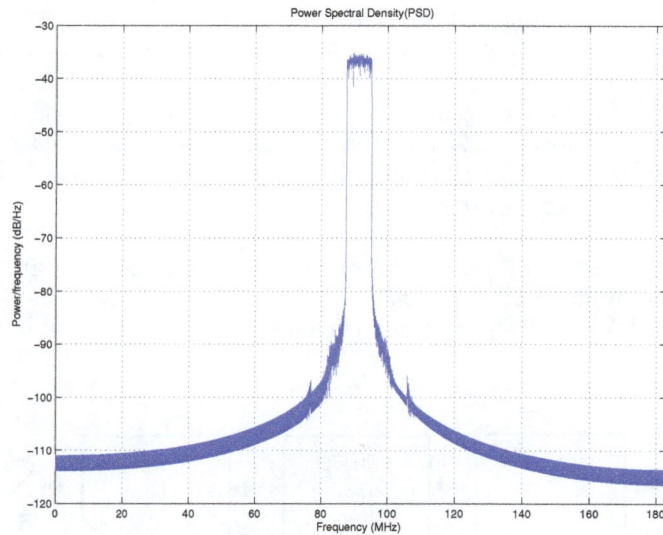

Figure 3. The Power Spectral Density (PSD) of transmitted symbols of DVB.

6. CLASSIC FILTERS

The Butterworth filter has drawn considerable attentions in communication system design. This type of filter exhibits a nearly flat pass band with no ripple in it. The roll off of Butterworth filter is very smooth and monotonic. For a low pass Butterworth filter the roll off rate is 20dB/decade for each added pole in the transfer function. In order to generate the spectrum of different filters we used MATLAB built-in function [36-39]. In MATLAB all digital frequencies are given in units of π. Hence, the ranges of the cut-off frequencies are defined by $0 \leq w_c \leq \pi$. In the rest of this paper we denote the cut-off frequency as $w_c = 1/2, 1/3, 1/4$, and $1/5$, where $w_c = 1/2$ is the highest cut-off frequency and $w_c = 1/5$ is the lowest cut-off frequency. We selected this frequency arbitrarily for this investigation. The spectrums of Butterworth filters of different orders are shown in Figure 4. In general a higher order filter has higher attenuation rate and vice versa as depicted in the figure. Another important characteristic of Butterworth filter is the cut-off frequency. The spectrums of Butterworth filter with different cut-off frequencies and the same order N=3 are shown in Figure 5. The filters with higher cut-off frequencies have higher attenuation rates as shown in the same figure. The second category filter that is considered in this investigation is the Chebyshev filter. The Chebyshev filter response has a faster roll-off. But, it has ripple in the frequency response. There is a trade-off between ripple and roll-off characteristic of this type of filter. As the ripple increases, the roll-off becomes higher. Like Butterworth filter the spectrums of Chebyshev filter also vary with respect to the order and cut-off frequency. The Chebyshev filters can be classified as Type-I and Type-II. Type-I Chebyshev filter has ripple in the pass band. On the other hand Type-II Chebyshev filter has ripple in the stop band. Since, Type-II filter is rarely used, we consider only Type-I filter in this investigation. Figure 6 shows the spectrums of Chebyshev filter of different order with a pass band ripple of 20 dB and cut-off frequencies of 180 MHz. Figure 7 shows the spectrums of Chebyshev filters with different cut-off frequencies. Compared to Butterworth filter the Chebyshev filter can achieve a sharper transition between the pass band and the stop band with a lower order filter. The sharp transition between the pass band and the stop band of a Chebyshev filter produces smaller absolute errors and faster execution speeds than a Butterworth filter.

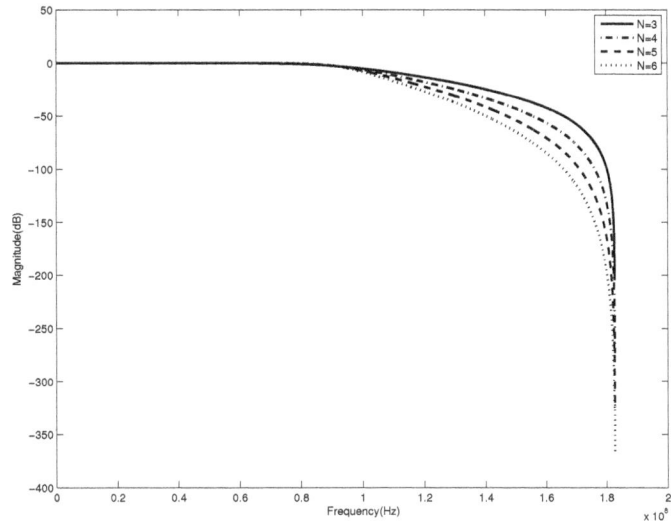

Figure 4. Spectrums Butterworth filters of different orders

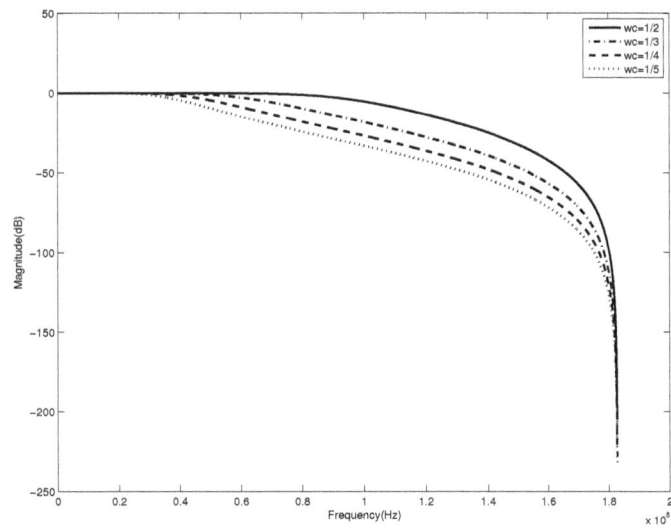

Figure 5. Spectrums of Butterworth filters of different cut-off frequencies

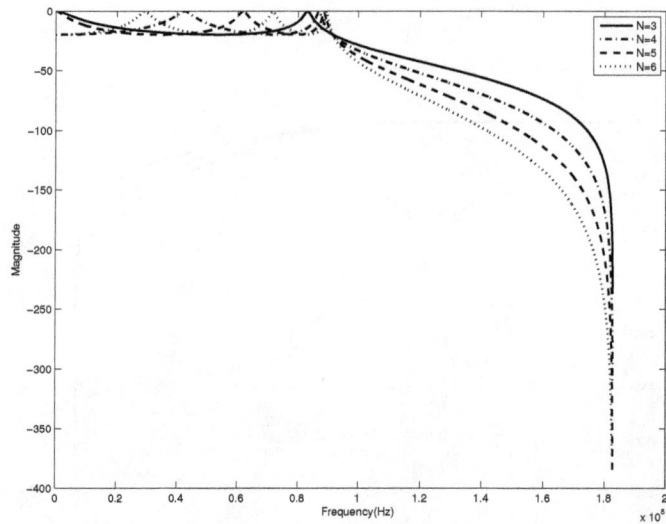

Figure 6. Spectrums of Chebyshev filters of different order

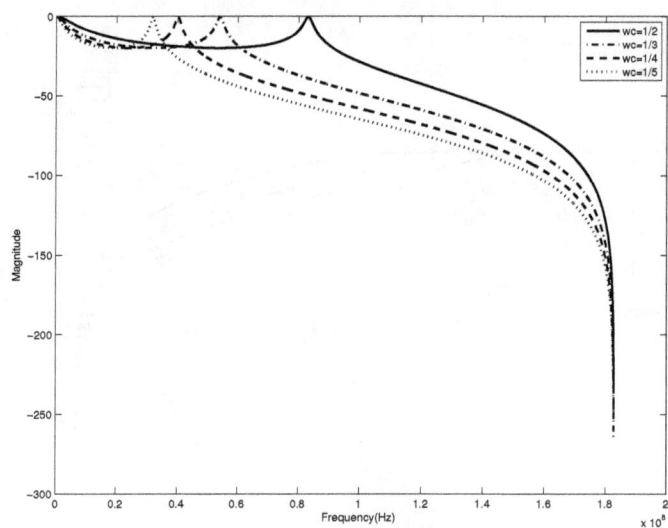

Figure 7. Spectrums of Chebyshev filters of different cut-off frequencies

The last category filter that has been considered in this investigation is the elliptic filter. The cut-off slope of an elliptic filter is steeper than that of a Butterworth and Chebyshev. But, this type of filter has ripple in the stop band. The phase response is very nonlinear. The spectrums of elliptic filters of different orders and cut-off frequencies are shown in Figure 8 and Figure 9 respectively. Compared with the same order Butterworth and Chebyshev filters, the elliptic filters provide the sharpest transition between the pass band and the stop band.

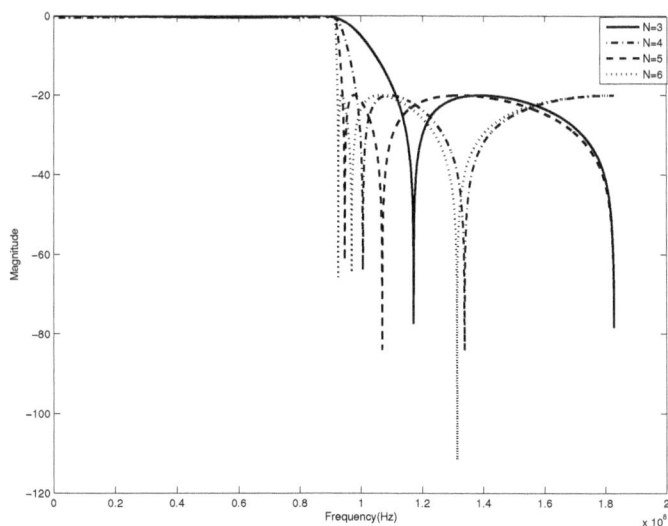

Figure 8. Spectrums of Elliptic filters of different order

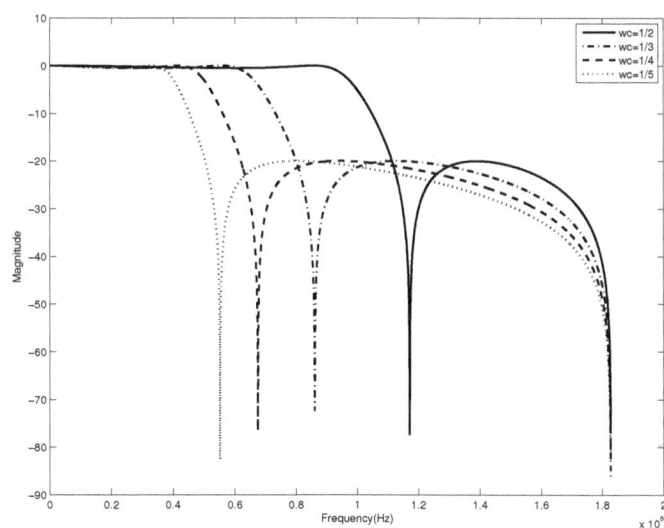

Figure 9. Spectrums of Elliptic filters of different cut-off frequencies

7. DVB RECEIVER PERFORMANCES USING CLASSIC FILTERS

Since the effects of digital filters are the main focus in this investigation, the cyclic prefix was set to 0 in all simulations. The other parameters were set to the values as mentioned in Table 1 and Table 2. The channel model was chosen Additive White Gaussian Noise (AWGN).

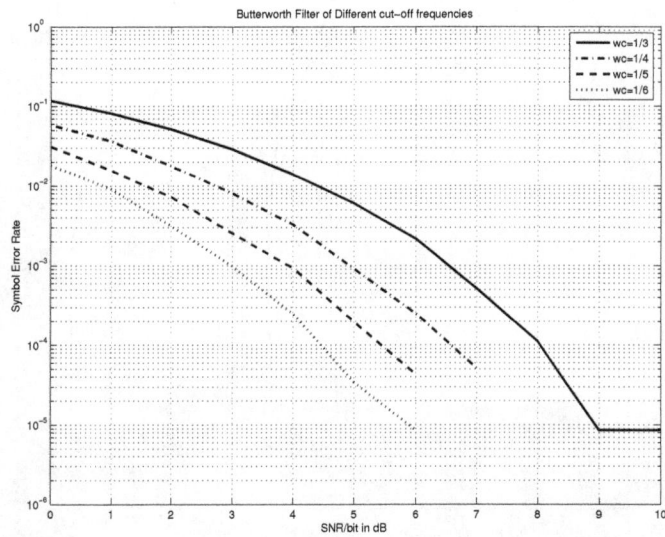

Figure 10. The SER error performances of OFDM system using Butterworth filter (cut-off frequency varied)

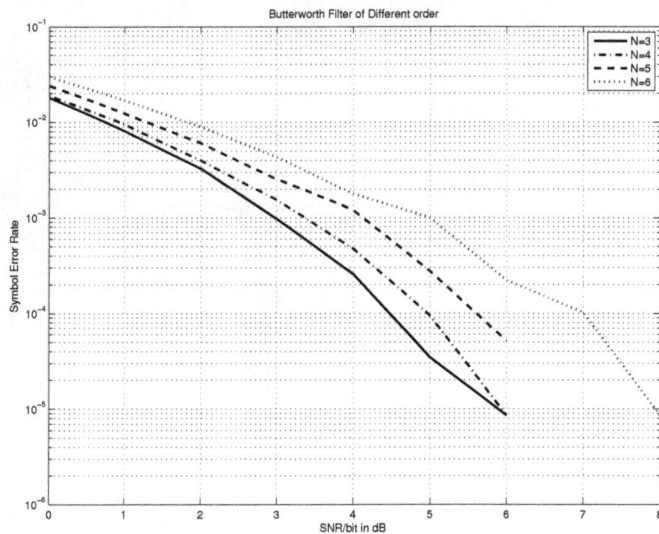

Figure 11. The SER performances of OFDM system using Butterworth filter (order varied)

In the first investigation we consider Butterworth filter in the receiver. The filter order was 3, but the cut-off frequency was varied. The simulation results are compared in Figure 10. The simulation results show that the lower cut-off frequency shows better performance (i.e. less symbol error rate) than the higher cut-off frequencies. For example, the results show that for a given SNR of 4 dB, the symbol error rate (SER) are 0.091, 0.005, 0.001, and 0.0009 for cut-off frequency of 1/3, 1/ 4,1/5 and 1/6 respectively. To investigate the performance of OFDM system under varying filter orders we chose a Butterworth filters with cut-off frequency of $w_c=1/5$ and we varied the filter order. The simulation results are shown in Figure 11. The simulation results presented therein show that the lower order Butterworth filters are preferable in DVB-T receiver. For a given SNR filter with lower order provides less SER. But, the DVB receiver performance is more dependent on the cut-off frequency other than filter order as illustrated in Figure 10 and Figure 11.

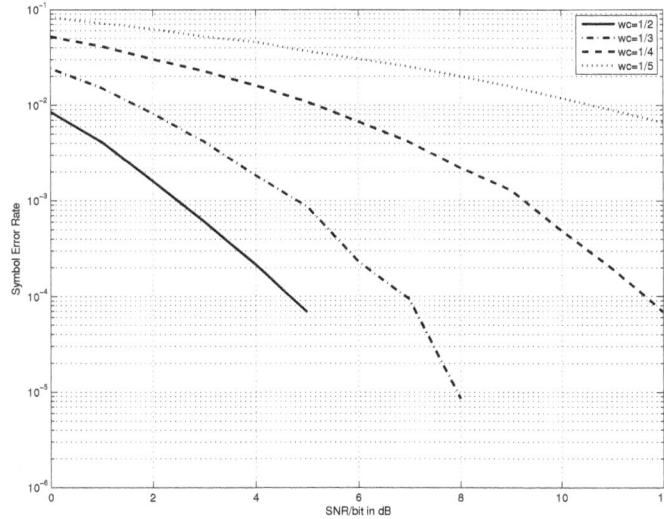

Figure 12. The SER performances of OFDM system using Chebyshev filter (cut-off frequency varied)

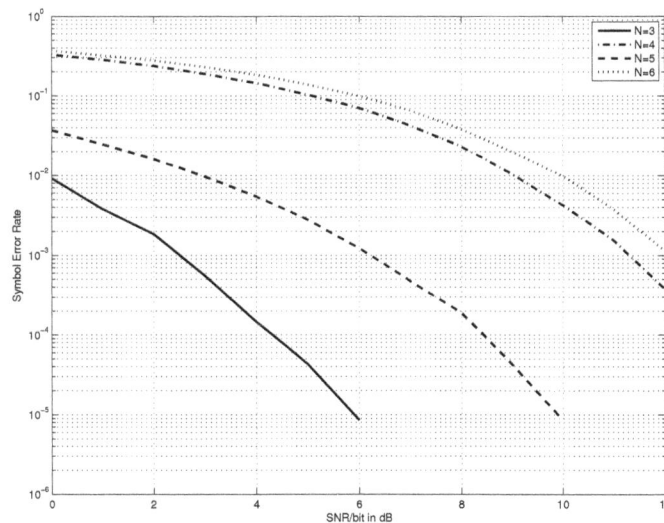

Figure 13. The BER performances of OFDM system using Chebyshev Filter (order was varied)

In the second investigation the receiver filter was then changed to Chebyshev filter by keeping all other parameters same. The filter order was set to 3 and the cut-off frequency was varied. The pass band ripple was set to 10 dB. The simulation results are shown in Figure 12. In contrast to Butterworth filter the simulation results show that Chebyshev filters with higher cut-off frequencies are desirable for DVB-T receiver. For a given SNR Chebyshev filter caused less SER compared to Butterworth filter. For an example, the results show that when SNR was set to 4 dB, the symbol error rate (SER) are 0.04, 0.02, 0.002, and 0.0002 for cut-off frequencies 1/3, 1/ 4,1/5 and 1/6 respectively. Compared to those of Butterworth filter it is evident that SER is less in Chebyshev filter. For an example, In the next investigation the filter cut-off frequency was kept constant at 1/2, but the filter order was varied. The simulation results are shown in Figure 13. The results show that a low filter order performs better than a higher order filter. The performance variation is similar to that of Butterworth filter.

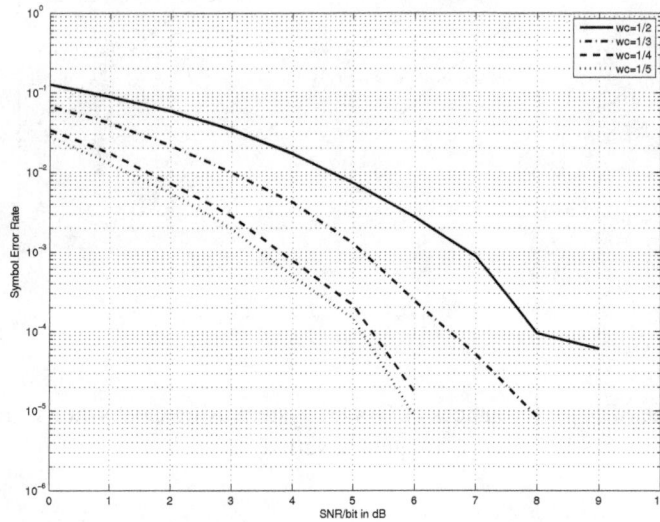

Figure 14. The BER performances of OFDM system   elliptic filter (cut-off frequency varied)

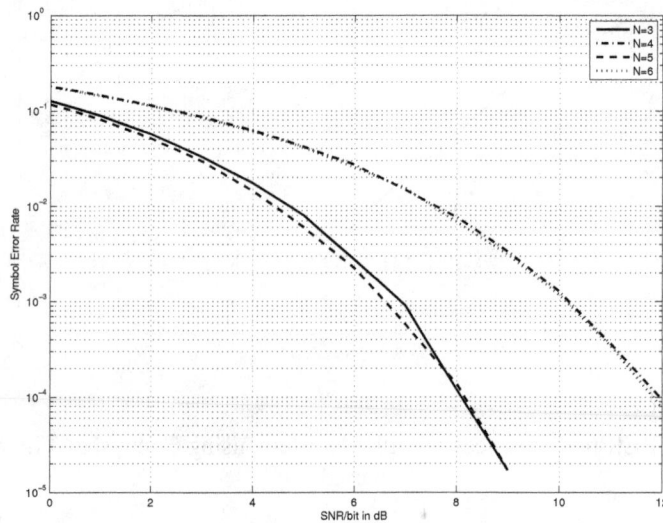

Figure 15. The BER performances of OFDM system using Chebyshev (order varied).

In the last investigation the Chebyshev filter is replaced by an elliptic filter by keeping other parameters same. We selected an elliptic filter with order 3. The pass band ripple was maintained at 0.5 dB and the stop band ripple was set to 10 dB. The cut-off frequency was varied to different values. The simulation results are shown in Figure 14. It is depicted in the figure that the elliptic filter performs almost in the same way as the Butterworth filters does. For example, at the SNR of 4 dB, the symbol error rates are 0.02, 0.002, 0.0004 and 0.0003 for cut-off frequencies of 1/2, 1 /3, 1/4, and 1/5 respectively. The figure also depicts that the elliptic filters with lower cut-off frequencies perform better than those of higher cut-off frequencies. Then the filter order was varied and the cut-off frequency was fixed to ½. The simulation results are shown in Figure 15. It is shown that a higher order filter is desirable for OFDM receiver. But, the variation of the performances with the filter order is not significant in all cases.

8. CONCLUSIONS

In this paper the performance of OFDM based DVB-T receiver has been investigated. OFDM modulation scheme has drawn wide attentions by the researchers in academia and industry. Designing a good OFDM receiver design is still an open issue. It is shown in this paper that OFDM receiver performance depends heavily on the filter types. Different filters have been taken into account. It is shown that the filter types and some important filter parameters like filter order and cut-off frequency affect the OFDM receiver performance. It is widely accepted that the Butterworth filter is the most popular choice for OFDM receiver. But, Chebyshev filters and elliptic filters are also good candidate. It is also shown that OFDM receiver is more sensitive to Chebyshev and elliptic filters other than Butterworth filters.

REFERENCES

[1]   Wu. Y, Hirakawa S., Reimers, U.H. and Whitaker, J., "Overview of Digital Television Development Worldwide", In the Proceedings of IEEE, Vol. 94, January 2006, pp. 8-21

[2]   Udelson, J.H., "The Great Television Race: A History of the American Television Industry", University of Alabama Press, 1982

[3]   "The World in 2011: ICT Facts and Figures", Technical Report of International Telecommunication Union (ITU), October 2011.

[4]   "Television History " available at http://www.tvhostory.tv

[5]   www.dvb.org

[6]   Peter, S., "DVB: Developing Global Television Standards for Today and Tomorrow", Technical Symposium at ITU , 2011

[7]   Reimars, U.H., "DVB-The Family of International Standards for Digital Video Broadcasting", In the Proceedings of IEEE, Vol. 94, No.1, pp. 173-182

[8]   Ladebusch, U. and Liss, C.A., "Terrestrial DVB (DVB-T): A Broadcast Technology for Stationary Portable and Mobile Use", In the Proceedings of IEEE, Vol. 94, No.1, pp. 183-193

[9]   Introduction to the DVB Project, DVB Fact Sheet, May 2013 available at

http://www.dvb.org/technology/fact_sheets/DVB-Project_Factsheet.pdf

[10]   ETSI EN 300 744, Digital Video Broadcasting (DVB): Measurement Guidelines for DVB system, v. 1.2.1 (2009-01) available at http://www.etsi.org/deliver/etsi_en/ 300700_300799/ 300744/ 01.06.01_60/ en_300744v010601p.pdf

[11]   Popovic, M. L.; Sunjevaric, M. M., and Vujasinovic, Z.J. , "Effects of DVB-H and 3G cellular networks integration ", In the Proceedings of the 10th International Conference on Telecommunications in Modern Satellite Cable and Broadcasting Services, Serbia, October 2011, pp. 641-644

[12]   Nee, R.V. and Prasad, R. "OFDM Wireless Multimedia Communications", Norwood, MA: Artech House, 2000

[13]   Bingham, J.A.C. , "Multi-carrier modulation for data transmission: An idea whose time has come", IEEE Communications Magazine, Vol.28, No. 5, May 1990, pp. 5-14

[14]   Pun,M.O., Morelli, M. and Kuo, J., "Multi-Carrier Techniques for Broadband Wireless Communications: A Signal Processing Perspective," Imperial College Press, 57 Shelton Street, Covent Garden, London WC2H 9HE

[15]   Hirosaki, B. "An Analysis of automatic equalizer for orthogonally multiplexed QAM system", IEEE Transaction on Communication, Vol. com 28, No. 1, 1980, pp. 73-83

[16]   Vahlin, N., and Holte, N., " Optimal finite duration pulses for OFDM", IEEE Transaction on Communications, Vol. 44, No. 11, January 1996, pp. 10-14

[17]   Slepian, D. and Pollak,H., " Prolate speroidal wave functions , Fourier analysis and uncertainty-I", Bell System Technical Journal, Vol. 40, January 1961, pp. 43-63

[18]   Chen,H.H., and CAI, X.D., " Optimization of Transmitter and Receiver Filters for OQAM-OFDM Systems using Non-linear Programming", IEEE Transaction on Communications, Vol. E80-B, No. 11, November 1997, pp. 1680-1687

[19]   Aldis, J.P., Althoff, M.P. and Van-Nee, R., " Physical layer architecture and performance in the WAND user trial system", In the proceedings of ACTS Mobile Conference, Grenada, Spain, 1996, pp. 196-203

[20]   Gudmundson, M. and Anderson, P.O., "Adjacent channel interference in an OFDM system", In the Proceedings of IEEE Vehicular Technology Conference, 1996, pp. 918-922

[21]   Cimini, L.J., "Analysis and simulation of a digital mobile channel using orthogonal frequency division multiplexing", IEEE Transaction on Communications, Vol. 33, 1985, pp. 665-675

[22]   Khalid Aslam, Bodiuzzaman,M., Uddin J., and Kulesza, W, "Using Raised Cosine Filter to Reduce Inter- Symbol Interference in OFDM with BPSK technique", Computer Science Journal, Vol. 1, No. 2, August 2011, pp. 115-119

[23]   Faulkner, M., " The effects of Filtering on the performance of OFDM systems", IEEE Transactions on Vehicular Technology, Vol. 49, No.5, September 2000, pp. 1877-1884

[24]   Prafullah D. G. and Siddharth A. L., " PAPR Performance of OFDM system by using Clipping and Filtering", International Journal of Advances in Engineering and Technology, Vol. 6, No. 2, May 2013, pp. 789-794

[25]   Wang , L. and Tallanbara, C., "A simplified clipping and Filtering technique for PAR Reduction in OFDM systems", IEEE Signal Processing Letters, Vol. 12, No. 6, June 2005, pp. 453-456

[26]   Zhong, X., Qi. J, and Bao, J., " Using clipping and filtering algorithm to reduce PAPR of OFDM system", In the Proceedings of the International Conference on Electronics, Communication, and Control, September 2011, Ningbo, China, pp. 1763-1766

[27]   Aziz W., Ahmed E., Abbas G., Saleem, S., and Islam, Q. " PAPR Reduction in OFDM using Clipping and Filtering" , World Applied Sciences Journal, January 2012, Vol. 11, pp. 1495-1500

[28]   Ho C.K., Sun S., and Farhang B., " Detrimental Effects of filtering in an OFDM system using pilot based channel estimation", In the Proceedings of the 13th IEEE International Symposium on Personal, Indoor, and Mobile Radio Communication, September , 2002, Vol. 3, pp. 1316-1320

[29]   Naveen V,.J., and Rajeswan K.R, " ICI Reduction using Extended Kalman Filter in OFDM system", International Journal of Computer Application, Vol. 17, No. 7, March 2011, pp. 15-22

[30]   Li X., and Cimini L.J, "Effects of clipping and filtering on the performance of OFDM", IEEE Communication Letter, Vol. 2, No.5, May 1998, pp. 131-138

[31]   Klinler, F. and Scelze, H., " Adjacent Channel Interference with Imperfect Anti-Aliasing Filtering", In the Proceedings of the 12th International OFDM- Workshop , August 07, Hamburg, Germany

[32]   Saritha , H.M. and Kulakarni, M. , " A novel scheme using MLE with pulse shaping for ICI cancellation in OFDM system", In the Proceedings of International Conference on Communication and Informatics, January 2012, Coimbatore, India, pp. 1-5.

[33]   Chang, S. and Powers E.J., "Cancellation of inter-carrier interference in OFDM systems using a nonlinear adaptive filter", In the Proceedings of IEEE International Conference on Communication, 2000, Vol. 2, pp. 1039-1043

[34]   OFDM Simulation using Matlab available at http://www.ece.gatech.edu/research/_labs/ sarl/ tutorials / OFDM/Tutorial_web.pdf

[35]   John G. Proakis and Dimitris G. Manolakis, "Digital Signal Processing : Principles, Algorithms, and Applications", Fourth Edition, Prentice Hall, Upper Saddle River, New Jersey, USA

[36]   Ingle, V.K. and Proakis J.G., "Digital Signal Processing using MATLAB", Thomson Bookware Companion Series

[37]   Antoniou A., "Digital Signal Processing: Signals, Systems and Filters", McGraw Hill, New York, USA

[38]   Mitra S.K., "Digital Signal Processing: A Computer Base Approach", McGraw Hill, New York, USA

[39]   Bose T., "Digital Signal and Image Processing", John Wiley and Sons, New Jersey, USA

Effect of Node Mobility on AOMDV Protocol in MANET

Indrani Das[1], D.K Lobiyal[2] and C.P Katti[3]

[1,2,3]School of Computer and Systems Sciences
Jawaharlal Nehru University, New Delhi, India.

ABSRACT

In this paper, we have analyzed the effect of node mobility on theperformance of AOMDV multipath routing protocol. This routing protocol in ad hoc network has been analyzed with random way point mobility model only. This is not sufficient to evaluate the behavior of a routing protocol. Therefore, in this paper, we have considered Random waypoint, Random Direction and Probabilistic Random Walk mobility Model for performance analysis of AOMDV protocol. The result reveals that packet delivery ratio decreases with the increasing node mobility forall mobility models. Also, average end-to-end delay is also vary with varying node speed, initially upto 20 nodes in all mobility models delay is minimum.

KEYWORDS

AOMDV, Multipath Routing, Ad hoc Network, Packet delivery ratio, Average end-to-end delay,Mobility models.

1.INTRODUCTION

A Mobile Ad-Hoc Network (MANET) is a network where more than two autonomous mobile hosts (mobile devices i.e. mobile phone, laptop, iPod, PDAs etc.) can communicate to each other without any mean of fixed infrastructure.When source(S) node want to send some data toward the destination (D), if they are fall in the same transmission range only can directly communicate with each other.Otherwise with the help of intermediate nodes communication can be established. Any node may join and leave the network in any point of time, therefore the topology of the network changes frequently. In this network some scarce resources like battery power of mobile devices, bandwidth of network. The depletion of battery power may affect lifetime of the whole network as well as individual node existence in the network.Due to dynamic topology and other network constraint routing in MANET is a challenging issue. Single path routing is not always sufficient to disseminate data to the destination. Therefore; multipath routing comes into existence toovercome the problem of single path routing.

In this paper we have considered various mobility models for proper and in depth analysis of AOMDV protocol. In literature we have discussed various works related to AOMDV protocol and brief about various multipath routing protocols. Most of the work carried out based on random waypoint mobility model.Therefore,we have analyzed AOMDV protocol with various network parameters and mobility models. Finally, we have computed packet delivery ratio and average end-to-end delay with varying node speed for individual mobility models.

The rest of the paper is organized as follows. In section II we have discussed various works related to multipath routing. In section III, various mobility models and AOMDV routing protocolbriefly discussed. Results analysis and simulation work is presented in Section IV and finally, we have concluded the paper in Section V.

2. RELATED WORKS

Multipath routing overcomes various problems occurs while data delivered through a single path. The multipath routing protocols are broadly classified based on on-demand, table driven, and hybrid. The following multipath routing protocols are used in MANETs. In [1] authors have compared the performance of AOMDV and OLSR routing protocol with Levy-Walk and Gauss-Markov Mobility Model. For the analysis they have considered varying mobility speed and the traffic load in the network. Their results show that AOMDV protocol achieved higher packet delivery ratio and throughput compared to OLSR. Further, OLSR has less delay and routing overhead at varying node density. In [2] authors only compared AOMDV and AODV routing protocol with random way point mobility model. Different traffic source like TCP and CBR is considered. The result shows that with increasing traffic both routing protocols performance degraded. In M-DSR (Multipath Dynamic Source Routing) [5, 21] is an on demand routing protocol based on DSR [12]actually it is a multipath extension of DSR. In SMR (Split Multipath Routing) [5, 15] is an on demand routing protocol and extension of well- known DSR protocol. The main aim of this protocol is to split the traffic into multiple paths so that bandwidth utilization goes in an efficient manner. In GMR (Graph based Multipath Routing) [5, 9] protocol based on DSR, a destination node compute disjoint path in the network using network topology graph.In MP-DSR [5, 13, 16] is based on DSR; it is design to improve QoS support with respect to end-to-end delay. In [10,19] authors have proposed an on-demand multipath routing protocol AODV-BR. But to establish multipath it does not spend extra control message. This protocol utilizes mesh structure to provide multiple alternate paths. In [8] authors have considered node-disjoint and link-disjoint multi-path routing protocol for their analysis. The various mobility models Random Waypoint, Random Direction, Gauss-Markov, City Section and Manhattanmodels are considered. Through the thorough analysis they have shown that in Gauss markov mobility model multipath formation is less but path stability is high. (The random direction model form larger number of multipath.) In [14] authors have considered AODV and AOMDV protocol for their performance analysis with random waypoint model. The result shows that AOMDV has more routing overhead and average end to end delay compared to AODV. But AOMDV perform better in term of packets drops and packet delivery. In [17] various energy models with Random Waypoint Mobility Model,Steady State mobility model is used to analyze the energy overhead in AOMDV, TORA and OLSR routing protocols. Results show that TORA protocol has highest energy overhead in all the energy models.In [22] performance of AOMDV protocol is analyzedfor different mobility models to investigate how this protocol behaves in different mobility scenario. The results show that with increasing node density, packet delivery ratio increases but with increasing node mobility packet delivery ratio decreases.

3. DESCRIPTION OF ROUTING PROTOCOL AND MOBILITY MODELS

In this section we have discussed brief about AOMDV routing protocol and various mobility models considered for simulation work.

3.1 Ad Hoc On Demand Multipath Distance Vector (AOMDV)

Ad Hoc On Demand Multipath Distance Vector (AOMDV) [3, 5, 6, 11] protocol is a multipath variation of AODV protocol. The main objective is to achieve efficient fault tolerance i.e. quickly recovery from route failure. The protocol computes multiple link disjoint loop free paths per route discovery. If one path fails the protocol choose alternate route from other available paths. The route discovery process is initiated only when to a particular destination fails. When a source needs a route to destination will floods the RREQ for the destination and at the intermediate nodes all duplicate RREQ are examined and each RREQ packet define an alternate route. However, only link disjoint routes are selected (node disjoint routes are also link disjoint). The destination node replies only k copies of out of many link disjoint path, i.e. RREQ packets arrive through unique neighbors, apart from the first hop are replied. Further, to avoid loop 'advertised hop count' is used in the routing table of node .The protocol only accepts alternate route with hop count less than the advertised hop count. A node can receive a routing update via a RREQ or RREP packet either forming or updating a forward or reverse path .Such routing updates received via RREQ and RREP as routing advertisement.

3.2 Mobility Models

Mobility pattern of node plays a vital role in evaluation of any routing protocol in MANET. We have considered various categories mobility models for acceptability of routing protocol. The following mobility model we have considered in simulation work.

3.2.1Random Waypoint Model

The Random Waypoint (RWP) mobility model [4,7] is the only model which is used in maximum cases for evaluation of MANET routing protocols. In this model nodes movement depends on mobility speed, and pause time. Nodes are moving in a plane and choose a new destination according to their speed. Pause time indicate that a node to wait in a position before moved to new position.

3.2.2 Probabilistic Random Walk Model

In this model [4,7]nodes next position is determined by set of probabilities. A node can be move forward, backward or remain in x and y direction depends on the probability defined in probability matrix. There are three state of node is defined by 0 (current position), 1 (previous position) and 2 (next position). Where, in the matrix P (a,b) means the probability that an node will move from state a to state b.

3.2.3 Random Direction Model

The random direction model [4,7] is the further modification of Random waypoint mobility model.This model overcome the density wave problem occur in random waypoint model, where clustering of nodes occur in a particular area of simulation. In Random Waypoint model this density occurs in the center of the simulation area. Here, nodes are move upto the boundary of the simulation area before moving to a new location with new speed and direction. When nodes are reached to the boundary of simulation area, before changing to new position it pauses there for sometimes. The random direction it chooses from 0 to 180 degrees. The same process is continued till the simulation time.

4. SIMULATION SETUP AND RESULT ANALYSIS

For the simulation works we have used Bonn-Motion mobility generator [18] to generate the mobility of nodes based on various mobility models. The most popular network simulator NS-2.34 [20] has beenused for simulation. Finally, in table-1 and table-2 different simulation parameters and their values have been shown respectively..

Table 1. Simulation Parameters

Parameter	Specifications
MAC Protocol	IEEE 802.11
Routing Protocol	AOMDV
Radio Propagation Model	Two-ray ground reflection model
Channel type	Wireless channel
Antenna model	Omni-directional
Mobility Models	Random Waypoint, Random Direction, Probabilistic Random Walk

Table 2. Values of Simulation Parameters

Parameter	Values
Simulation Time	1000s
Simulation Area (X *Y)	1000 m x1000 m
Transmission Range	250 m
Bandwidth	2 Mbps
No. of Nodes	10,20,30,40,50,60,70,80.90, 100
Node speed	10,20,30,40 m/s

Figure1 shows Packet delivery ratio with Random Waypoint Mobility Model for different node speeds. In this model AOMDV gives better packet delivery ratio with increasing node density. But with the increasing node mobility PDR decreases due to frequent link breakage among nodes. The maximum achievable value of PDR is 77.8%.

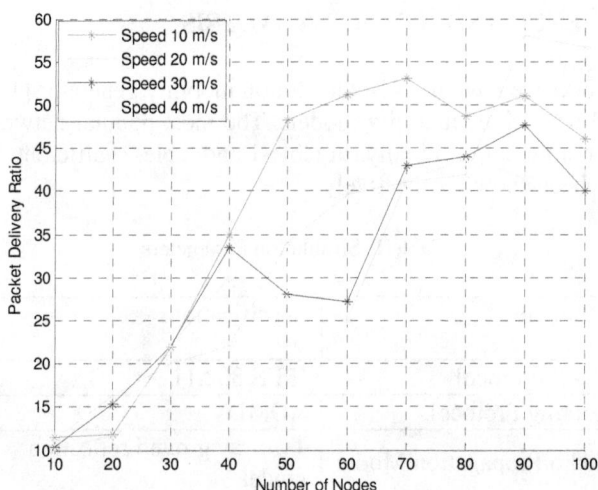

Figure1.Packet delivery ratio with variousnode speed (Random Waypoint Mobility Model).

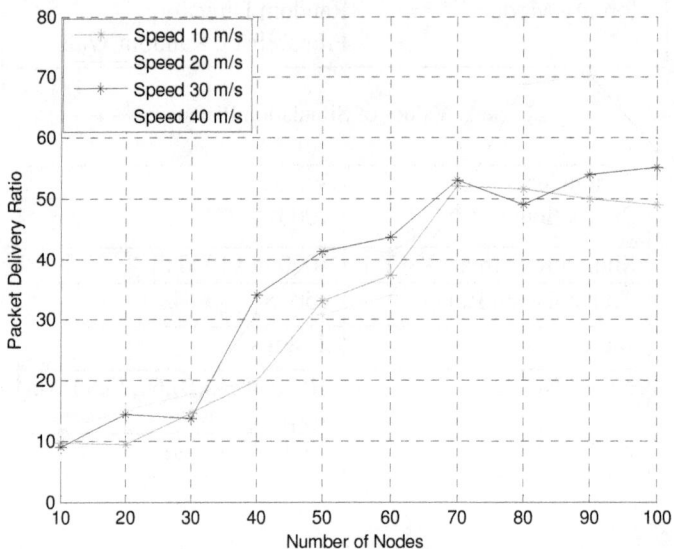

Figure2. Packet delivery ratio with variable node speed(Random Direction Mobility Model).

Figure.2 shows the packet delivery ratio for differentnode mobility for Random Direction Mobility model. In this mobility model PDR decreases with the increasing node mobility. For 90 nodes and speed of 20 m/s maximum value of PDR i.e. 70% is achieved. There is a sudden drop in the PDR as the number of nodes increases beyond 90 due to congestion, sudden link failure etc.At node speed 10 and 20 m/s protocol performance is steady and with increasing speed, the value of PDR reduces.

Figure.3 shows the packet delivery ratio for different node mobility using Probabilistic Random walk mobility model. In this modelAOMDV gives poor performance in term of PDR when node speed increases due to frequent link breakage among nodes. The maximum achievable value of PDR is 80% approximatelyat thenode speed of 10 m/s.

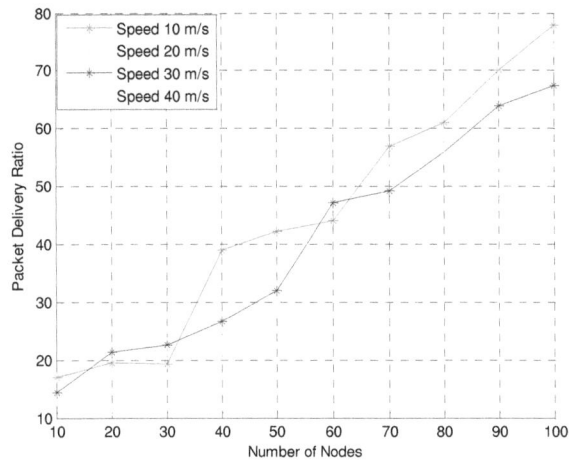

Figure3. Packet delivery ratio with variable node speed (Probabilistic Random Walk Model).

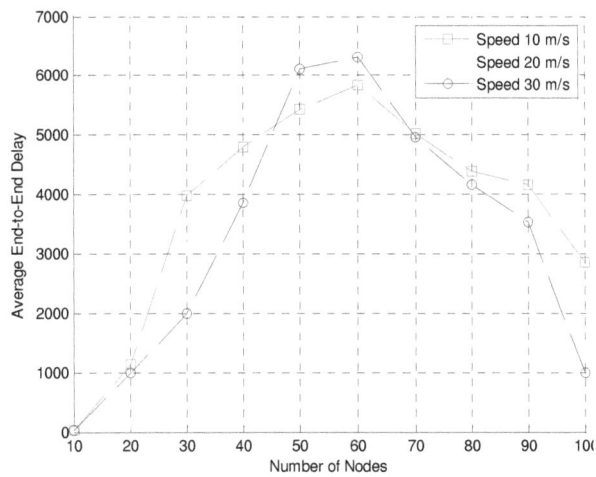

Figure 4.Average end-to-end delay with variable node speed (Random Waypoint Mobility Model).

Figure.4 shows the Average end-to-end delay with variable node speed in random waypoint mobility model. Here, upto 20 nodes in different speed delay is minimum but increases gradually as number of node increases. The maximum delay noticed when speed is 30 m/s and number of node 60. The delay is gradually decreases when number of nodes more than 60 onwards.

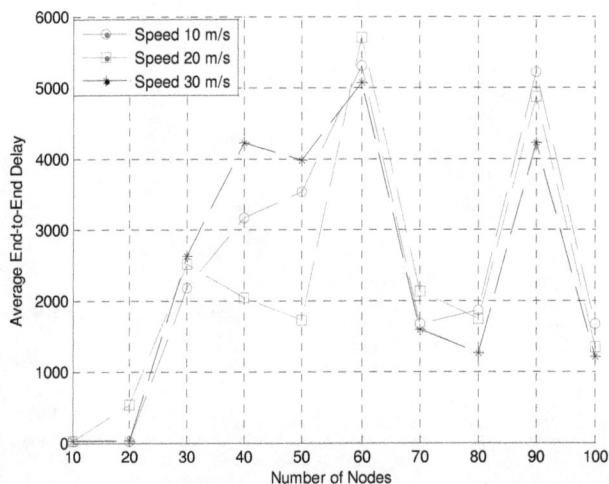

Figure 5.Average end-to-end delay with variable node speed (Random Direction Mobility Model).

Figure.5 shows the Average end-to-end delay with variable node speed in random direction mobility model.Here, upto node 20 in different speed delay is minimum but increases gradually as node increases. The maximum delay noticed when speed is 20 m/s and number of node 60. Further, sudden fall in delay is noticed at number of node is 70 and 80.

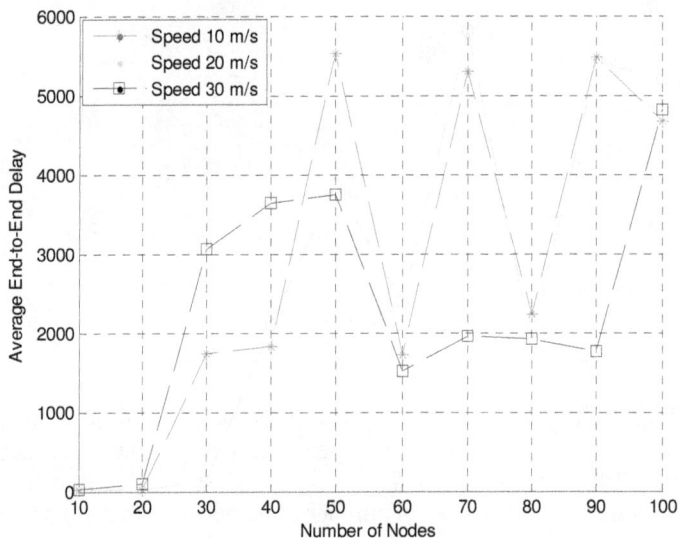

Figure 6.Average end-to-end delay with variable node speed (Probabilistic Random Walk Model).

Figure.6 shows the Average end-to-end delay in probabilistic random walk model with variable node speed. Here, upto 20 nodes in different node speed delay is minimum. As the speed and node vary delay gradually increases. At node speed 30 m/s from node 60 to 90 a consistent delay is noticed.

The overall analysis shows that with high node mobility the value of PDR decreasesfor all mobility models. The average end-to-end delay is gradually increases with increasing speed and nodes, but in random direction and random waypoint mobility models maximum delay noticedat 60 nodes. The delay is gradually decreases 60 nodes onwards. In probabilistic random walk model at node speed 30 m/s average delay in minimum as compare other speed.The overall performance of AOMDV protocol performs better for Randomwaypoint mobility model as compared to other mobility models.

5. Conclusions

We have evaluated the effect of node mobility on the performance of AOMDV multipath routing protocol with different mobility models. For the performance analysis of the protocol packet delivery ratio is computed. It is evident from the results that AOMDV protocol perform better in term of PDR and average end-to-end delay for Random Waypoint mobility model. But it is also noticed that with higher node mobility PDR of AOMDV protocol decreases. In Probabilistic Random walk model upto 70 nodes with various node mobility protocol performs better as compared to others. In future, this multipath protocol can be investigated for different network topologies.

References

[1] Gowrishankar. S, et al. (2010) "Analysis of AOMDV and OLSR Routing Protocols under Levy-Walk Mobility Model and Gauss-Markov Mobility Model for Ad Hoc Networks", (IJCSE) International Journal on Computer Science and Engineering, Vol. 02, No. 04, 2010, pp. 979-986.

[2] Vivek B. Kute et al., (2013) "Analysis of Quality of Service for the AOMDV Routing Protocol", ETASR - Engineering, Technology & Applied Science Research Vol. 3,No. 1, pp.359-362.

[3] JiaziYi ,AsmaaAdnane, Sylvain David, and BenoîtParrein, (2011) "Multipath optimized link state routing for mobile ad hoc networks", Ad Hoc Networks, Vol. 9, No.1, pp. 28-47.

[4] RadhikaRanjan Roy, (2011)Handbook of Mobile Ad Hoc Networks for Mobility Models, First Edition, Springer, New York Dordrecht Heidelberg London, ISBN 978-1-4419-6048-1 e-ISBN 978-1-4419-6050-4.

[5] Tsai, J., & Moors, T., (2006) "A review of multipath routing protocols: from wireless ad hoc to mesh networks", In Proceedings of ACoRN early career researcher workshop on wireless multi-hop networking, Sydney.

[6] M. K. Marina and S. R. Das, (2006) "Ad-hoc on-demand multi-path distance vector routing",Wireless Communication Mobile Computing, Vol. 6, No. 7, pp. 969–988.

[7] Camp, Tracy et al., (2002) "A Survey of Mobility Models for Ad Hoc Network Research", wireless communications & mobile computing (WCMC): special issue on mobile ad hoc networking: research, trends and applications, Vol.2, No.5,pp. 483-502.

[8] Nicholas cooper et al., (2010) "Impact of Mobility models on multipath routing in mobile Ad hoc Networks", International Journal Of Computer Networks & Communications (IJCNC), Vol. 2, No.1, pp.185-194.

[9] GunyoungKoh, Duyoung Oh and Heekyoung Woo, (2003) "A graph-based approach to compute multiple paths in mobile ad hoc networks", Lecture Notes in Computer Science Vol.2713, Springer, pp. 3201-3205.

[10] M.T.Toussaint, (2003)"Multipath Routing in Mobile Ad Hoc Networks", TU-Delft/TNO Traineeship Report.

[11] S. Das, C. Perkins and E. Royer (2003) "Ad Hoc On Demand Distance Vector (AODV) Routing", IETF RFC3561.

[12] D. Johnson, (2003)"The Dynamic Source Routing Protocol for Mobile Ad Hoc Networks(DSR)", IETF Internet Draft, draft-ietf-manet-dsr-09.txt.

[13] E. Esmaeili, P. Akhlaghi, M. Dehghan, M.Fathi,(2006) "A New Multi-Path Routing Algorithm with Local Recovery Capability in Mobile Ad hoc Networks", In the Proceeding of 5th International Symposium on Communication Systems, Networks and Digital Signal Processing (CSNDSP 2006), Patras, Greece, pp. 106-110.

[14] R.Balakrishna et al., (2010) "Performance issues on AODV and AOMDV for MANETS",International Journal of Computer Science and Information Technologies, Vol. 1, Issue.2, pp. 38-43.

[15]   S. J. Lee and M. Gerla, (2001) "Split Multipath Routing with Maximally Disjoint Paths in Ad Hoc Networks", In Proceedings of the IEEE ICC, pp. 3201-3205.

[16]   R. Leung, J. Liu, E. Poon, Ah-Lot. Chan, B. Li, (2001) "MP-DSR: A QoS-Aware Multi-Path Dynamic Source Routing Protocol for Wireless Ad-Hoc Networks", In Proc. of 26th Annual IEEE Conference on Local Computer Networks (LCN), pp. 132-141.

[17]   Gowrishankar.S et al., (2010) "Simulation Based Overhead Analysis of AOMDV, TORA and OLSR in MANET Using Various Energy Models", Proceedings of the World Congress on Engineering and Computer Science,San Francisco, USA, Vol.1.

[18]   Bonn Motion, http://net.cs.uni-bonn.de/wg/cs/applications/bonnmotion/

[19]   Sung-Ju Lee and Mario Gerla, (2000) "AODV-BR: Backup Routing in Ad hoc Networks", IEEE Conference on Wireless Communications and Networking Conference (WCNC- 2000),Vol.3, PP. 1311-1316.

[20]   The Network Simulator. http://www.isi.edu/nsnam/ns/.

[21]   A. Nasipuri and S. R. Das, (1999)"On-demand multipath routing for mobile ad hoc networks", In the Proceedings of Eight IEEE International Conference   on Computer Communications and Networks, Boston, MA, pp.64-70.

[22]   Indrani Das, D.K Lobiyal and C.P Katti (2014) "Effect of Mobility Models on the Performance of Multipath Routing Protocol in MANET", Published in the proceeding of fourth International Conference on Advances in Computing and Information Technology (ACITY 2014), Volume Editors: DhinaharanNagamalai, SundarapandianVaidyanathan, May 24~25, 2014, Delhi, India, pp.149-155.(ISBN : 978-1-921987-22-9 ,DOI : 10.5121/csit.2014.4516).

[23]   The Math Works: http://www.mathworks.com

13

ElGamal Signature for Content Distribution with Network Coding

Alireza Ghodratabadi[1] and Hashem Moradmand Ziyabar[2]

[1] EE institute, Malekashtar University of Technology, Tehran, Iran
[2] IRIB, Tehran, Iran

Abstract

Network coding is a slightly new forwarding technique which receives various applications in traditional computer networks, wireless sensor networks and peer-to-peer systems. However, network coding is inherently vulnerable to pollution attacks by malicious nodes in the network. If any fake node in the network spreads polluted packets, the pollution of packets will spread quickly since the output of (even an) honest node is corrupted if at least one of the incoming packets is corrupted. There have been adapted a few ordinary signature schemes to network coding that allows nodes to check the validity of a packet without decoding. In this paper, we propose a scheme uses ElGamal signature in network coding. Our scheme makes use of the linearity property of the packets in a coded system, and allows nodes to check the integrity of the packets received easily.

Keywords

network coding; crystallographic signature; homomorphic signature; ElGamal signature

1. Introduction

In recent years peer to peer networks are dynamic and each node in P2P networks is an information source. These nodes can come into and leave out the network instantly. Decentralized structure of these networks make it possible that malicious nodes disguise honest nodes and insert fake data in network. When network coding is applied, as nodes use their input to make a linear combination for each output, malicious packet spread even more. This way even honest nodes spread malicious packets. Therefore methods to detect these malicious packets seem to be necessary.

Recently there were also interests in applying network coding to distributed file systems. Many researchers have studied the usability and efficiency of network coding in peer to peer networks for distributing file and media [1], [2], [3]. Others were trying to use network coding in huge networks of computers and contribute it with peer to peer networks like freenet to update operating systems and software packages.

In traditional network model which is based on server-user model, server sends file according to user request. This model shows its inefficiency when the file size is large or the number of requests for a specific file is high, needing more bandwidth. Due to these reasons there has been a growing demand for peer to peer networks.

These networks are overcoming server-user networks; they have distributed structure and each user is considered as a server that can share network resources such as memory, processing power, and etc.

The best example for these kinds of networks is bitTorrent in which files are broken to smaller blocks. Whenever a node downloads a block that node is considered as its server. However these systems are being used now but due to some problems its performance is degraded considerably. For example synchronization always matters [4]. Synchronization means which block or blocks should be uploaded sooner or later. This problem will be more serious when rare blocks are placed on nodes with unreliable internet connection. To solve this problem [5] and [6] proposed to use network coding to enhance performance. Same as file is broken to blocks and every time that a user asks for the file, server send a random linear combination of its blocks to it. When the user receives the first block of the file it will be as the server for that block.

In this method each node (user) is able to restore the file only when it receives enough blocks to solve a linear set of equations. This scheme which is based on network coding omits the need for synchronization [5]. In [7] it has been shown mathematically that using this scheme, performance and reliability are also enhanced.

Here we are interested in security of distributed content schemes that use network coding. Our main considerations are simplicity and security performance. In every network that uses network coding avoiding malicious node. For example in the discussed peer to peer network with network coding, if a fake node imitate a valid node it can make a polluted block with correct coding coefficients and spread it over the network [8]. If there is not a mechanism for checking integrity of blocks it is possible for fake nodes to insert their packets into the network and other nodes who receive some polluted packet along with valid packets are not able to decode messages correctly.
In previous schemes where network coding was not used, data integrity was checked by means of hash-and-sign scheme. Source node exploited a hash function H(.) to compute hash H(X) for its data X. then it uses a digital signature method to sign it S(H(X)) and then send it. To defend against this attack many methods are proposed. [5] and [8] used homomorphic hash functions. In fact homomorphism enables nodes to sign every combination of messages without knowing their private key. In [9] exploiting SRC is suggested that is less complex than previous solutions but needs a secure channel for SRCs. Authors of [4] tried to omit secure channel so they defined a orthogonal space from messages. If received message is in this space it will be known as valid else it will be thrown away.

In this paper we have proposed a method for protecting packets from pollution attacks. We used cryptographic methods that exploit linear random coding. Data packets are considered as vector blocks. Packet contents are signed with ElGamal signature. In each receiver node, the validity of signature is being verified. If the signature is not valid that block will be omitted. Otherwise the node will use the block for making its output blocks. Our method have a larger overhead in comparison to previous signatures, but it is more robust to polluted node attacks.

The organization of paper is as follows. In section II, ElGamal signature is described briefly. Our proposed method is presented in section III along with a detailed example. Finally section IV concludes the paper.

2. ElGamal Signature

The proposed scheme is based on well-known ElGamal signature scheme proposed in 1984. Here we briefly describe this signature scheme. In this scheme when Alice wants to sign a message and send it to Bob, she chooses a prime number p, a generator g of a product field, F_p^*, and a random number x such that $0 < x < p-1$. She computes her public key as $y = g^x \bmod p$ and spreads public key vector (p, y, g). Alice also chooses a secret random number k and computes s and r as,

$$\begin{cases} r = g^k \bmod p \\ (k, p-1) = 1 \bmod p \end{cases} \quad \begin{cases} s = k^{-1}(m - xr) \bmod p - 1 \\ \quad m : \text{message} \end{cases}$$

The signature of message m, is (r,s). Receiving the message and its signature, Bob want to verify it. First he checks whether $1 \le r \le p-1$. If it does not hold he rejects the message, otherwise he checks the congruence $y^r r^s = g^m \bmod p$. He accepts the signature if this congruence holds too.
Proposed Scheme

Using ElGamal signature in peer-to-peer networks where network coding is implemented, enhances security by determining intentionally inserted errors. Just with these methods, peer-to-peer network can gain the benefits of network coding.

First, we describe linear network coding briefly. Let $G = (V, E)$ be a directed graph. Suppose a source node $s \in V$ sends a large file F to set $T \subseteq V$ of nodes in V. File F is consisted of k n-dimensional vectors $\mathbf{m}_1, ..., \mathbf{m}_k \in F^n$, where F^n is the n-dimensional vector space over a finite field F. So the file is consisted of k vectors as,

$$\begin{cases} \mathbf{m}_1 = (m_{1,1}, ..., m_{1,n}) \\ \mathbf{m}_2 = (m_{2,1}, ..., m_{2,n}) \\ \quad ... \\ \mathbf{m}_k = (m_{k,1}, ..., m_{k,n}) \end{cases}$$

Then the source creates *augmented vectors* of $\mathbf{m}_i = (m_{i,1}, ..., m_{i,n})$ by setting

$$\bar{\mathbf{m}}_i = (\overbrace{0, ..., 0, 1, 0, ..., 0}^{k}, m_{i,1}, ..., m_{i,n})$$

that is, the dimension of each augmented vector is $t = k + n$, and the first *m* entries are all zero except at *i*-th, where it is 1.

Assuming that $\mathbf{m}_2$ is being transmitted through the network. The source chooses a big prime number p and assumes g_1 to g_n as generators of product field, F_p^*. Also it assumes x as secret key and computes y_i's as

$y_i = (g_i)^x \bmod p$

Then the source distributes (p, g_i, y_i) as public key. For signing a message, it chooses a random secret number k so as k is prime to p-1, i.e. $(k, p-1) = 1$. Now it computes r_i's as;

$$r_i = (g_i)^k \bmod p$$

And sets r as product of r_i's and computes s too:

$$r = \prod_{i=1}^{n} r_i$$

$$s = k^{-1} \prod_{i=1}^{n} (m_{ji} - r_i) \bmod(p-1)$$

The signature is send as (r,s). At each node, receiver can check the validity of message by $r^s.y^r$; According to ElGamal signature, this term should be equal to $(g_1,...,g_n)^m \bmod p$. If the receiver node gains this, it verify the message otherwise the message is incorrect and probably fake and will be thrown away.

Table 1. Algorithm of Signature Scheme

Step	Action
System parameters	1- Selecting big prime number
	2- Selecting g_1 to g_n as generator of product field, F_p^*
Producing key	1- Selecting x as secret key
	2- Computing y_1 to y_n; $y_i = (g_i)^x \bmod p$
	3- Distribute y_i's, g_i and p as public key
Producing signature	1- Selecting random number k and computing r_i's so that (k,p-1)=1 and $$r_i = (g_i)^k \bmod p \text{ and}$$ $$r = \prod_{i=1}^{n} r_i$$
	2- Computing $s = k^{-1} \prod_{i=1}^{n} (m_{ji} - r_i) \bmod(p-1)$
	3- Sending (r,s) as signature
Validating signature	checking $(g_1,...,g_n)^m \bmod p \overset{?}{=} r^s.(y_1...y_n)^r \bmod p$

ElGamal signature is from hardness degree of discrete logarithm in product field, F_p^*. So unless discrete logarithm is broken, it will be secure.

2.1. An example

Here we give an example to make the method clear. Let m=(1,0,1,1) in binary format which is equal to 11 in decimal.

Assume p=11, so primitive roots g_i's are {2,6,7,8} and private key x=9 and

$$y_1 = (g_1)^9 \bmod 11 = 6$$
$$y_2 = (g_2)^9 \bmod 11 = 2$$
$$y_3 = (g_3)^9 \bmod 11 = 8$$
$$y_4 = (g_4)^9 \bmod 11 = 7$$

Therefore the public key is $\left\{ 11, \overbrace{2,6,7,8}^{g_i\text{'s}}, \overbrace{6,2,8,7}^{y_i\text{'s}} \right\}$. Since K have to prime to p-1, $k, p-1 = 1$, it should be selected form set $\{3,5,7,9\}$. Assume $K = 7$. Computing r_i's:

$$r_1 = (g_1)^7 \bmod 11 = 7$$
$$r_2 = (g_2)^7 \bmod 11 = 8$$
$$r_3 = (g_3)^7 \bmod 11 = 6$$
$$r_4 = (g_4)^7 \bmod 11 = 2$$

So $r = r_1'.r_2'.r_3'.r_4' = 672$ and $r \bmod 11 = 1$. Now to find k^{-1} we should find x so that $K.x \bmod 10 = 1$ which concludes to $k^{-1} = 3$.

S is;

$$s = k^{-1} \prod_{i=1}^{n} (m_{ji} - r_i) \bmod (p-1) =$$
$$= 3(0 - 9 \times 7)(1 - 9 \times 8)(0 - 9 \times 6)(1 - 9 \times 2) \bmod 10$$
$$= 3(-62)(-71)(-54)(-17) \bmod 10 = 12123108 \bmod 10$$
$$= 8$$

So the signature is (672,8). To check the signature, this equation is checked;

$$(g_1, ..., g_n)^m \bmod p \overset{?}{=} r^s.(y_1 ... y_n)^r \bmod p$$
$$(2 \times 6 \times 7 \times 8)^{11} \bmod 11 \overset{?}{=} 672^8.672^{672} \bmod 11$$
$$\overset{OK}{1 = 1}$$

Thus the equality means a valid signature.

3. CONCLUSIONS

Security problem is a main obstacle in the implementation of content distribution networks using random linear network coding. To tackle this problem, instead of trying to fit an existing signature scheme to network coding based systems, in this paper, we proposed a new signature scheme that is made specifically for such systems. We introduced a signature vector for each file distributed, and the signature can be used to easily check the integrity of all the packets received for this file. We have shown that the proposed scheme is as hard as the Discrete Logarithm problem, and the overhead of this scheme is negligible for a large file.

ACKNOWLEDGEMENTS

The authors would like to thank everyone, just everyone!

REFERENCES

[1]   T. Ho, M. Medard, M. Effros and D. Karger, "The benefits of coding over routing in a randomized setting," IEEE Symposium on Information Theory, 2003.

[2]   Z. Li and B. Li, "Network coding: the case of multiple unicast sessions," 42th Annual Allerton Conference on Communication, control and Computing, 2004.

[3]   D. S. Lun, M. Medard and R. Koetter, "Network coding for efficient wireless unicast," International Zurich Seminar on Communications, 2006.

[4]   F. Zhao, "Signature for Content Distribution with Network Coding," 2009.

[5]   S. Acedanski, S. Deb, M. Medard and R. Koetter, "How good is random linear coding based distributed network storage?," netcode, 2005.

[6]   P. Rodriguez and C. Gkantsidis, "Network Coding for large scale content distribution," IEEE INFOCOM, 2005.

[7]   C. Gkantsidis, J. Miller and P. Rodriguez, "Comprehensive view of a live network coding P2P system," ACM SIGCOMM/USENIX Internet Measurement Conference, 2006.

[8]   C. Rodriguez and G. P., "Cooperative security for network coding file distribution," IEEE INFOCOM, 2006.

[9]   M. Krohn, M. Freedman and D. Mazieres, "On-the-fly verification of rateless erasure codes for efficient content distribution," IEEE Symposium on Security and Privacy, 2004.

14

IMPACT OF CLIENT ANTENNA'S ROTATION ANGLE AND HEIGHT OF 5G WI-FI ACCESS POINT ON INDOOR AMOUNT OF FADING

Jehad Hamamra, Hassan El-Sallabi and Khalid Qaraqe
Electrical Engineering Department
Texas A&M University at Qatar

ABSTRACT

This paper investigates the impact of antenna rotation's angle at the receiver side and antenna height at transmitter side on radio channel's amount of fading. Amount of fading is considered as a measure of severity of fading conditions in radio channels. It indicates how severe the fading level relative to Rayleigh fading channel. The results give an input to optimize height of 5G Wi-Fi access point for better link performance for different antenna's rotation angles at receiver side. The investigation covers three different indoor environments with different multipath dispersion levels in delay and direction domains; lecture hall, corridor, and banquet hall.

KEYWORDS

Reconfiguring Antenna, Radio Channel, Amount of Fading, Channel Model

1. INTRODUCTION

The coming generations of Wi-Fi wireless networks 5G-Wi-Fi, IEEE 802.11ac [1] and WiGig (Millimeter Wave), IEEE 802.11ad are expected to provide extremely high data rate communication services. These technologies are mainly required to support advanced services such as efficient real-time video communication for users. Vast majority of mobile users usually spend most of their time in indoor environments. In fact, these indoor environments are miscellaneous and diverse. Communication system performance depends on radio channel characteristics of each indoor sceanario. The radio channel characteristics are function of distribution of scatterers, mobility in the channel at both ends of the connection link, bandwidth, frequency, and antenna's position of both transmitter and receiver and their radiation patterns as well.

The IEEE 802.11ac has a mandatory operation at the 5 GHz frequency range and bandwidth of 80 MHz with 160 MHz available optionally. The interference level near 5 GHz is less than that at around 2 GHz due to less usage of the bandwidth. The increase in throughput comes mainly from the increase in bandwidth to deliver high data rates of video applications to user terminals. The technology supports multi-user, multi-input, multi-output (MU-MIMO) scenarios and utilizes the 256 quadrature amplitude modulation (QAM). The standard group approved TGac Channel Model Addendum v12 [2,3], which are mainly modifications of the IEEE802.11n channel models [4] for wide bandwidth. The adopted models assume fixed tapped delay line channel models with a tap spacing of 5 ns and with some defined directional clusters. The directional clusters parameters are defined for angle of arrival and its angular spread, and angle of departure and its angular spread for every tap within its directional cluster.

It is important to study the impact of re-configurability of antenna pattern vertical and horizontal polarizations due to its rotation angle at client's side and antenna heights at access point side in different indoor environments that have different channel characteristics in terms of dispersion in delay and direction domains. The adopted channel model by IEEE 802.11ac, i.e., TGac does not provide the capability for such study since it has fixed delayed components that do not vary with mobile station variability. So, this work adopts physics-based model that derives each ray's parameters from environment geometry.

In this work, we present the impact of re-configurability antenna patterns due to its rotation at client side and height of access point in three different common indoor environments (lecture hall, corridor and banquet hall) on amount of fading.

2. RADIO CHANNEL MODEL

Radio wave propagation in an indoor environment is confined by the scatterers from different directions such as side walls, opposite walls, floor and ceiling. For indoor line of sight propagation, signals arrive at the receiver via direct path, signals bounce between opposite walls, side walls, ceiling and floor or any combinations between these different surfaces. This makes the signal arrive from almost all possible azimuthal angles with large directional dispersion in the vertical directional plane. These different rays would experience different antenna gains and losses as a function of angular information, azimuthal and co-elevation departure and arrival directions. The adopted channel model in this work is based on the physics of the specular reflection propagation mechanism in addition to the line of sight component. These propagation mechanisms are presented in terms of multi-rays that exhibit multi-dimensional parameters. This work models also the RF propagation in commonly shaped cubical indoor environment. This particular shape of indoor environment includes different propagation scenarios such as a corridor, office, lecture hall, convention center, etc. The tested model essentially presents similar features to the model presented in [5]. The simulated propagation characteristics are determined by the input parameters to the model and communications link setup. The input parameters to the channel model include operating frequency, system bandwidth, antenna's polarization and the heights of transmitter and receiver, antennas' field pattern, electrical properties of scatterers, etc. The communication link setup includes the locations of transmitter and receiver antennas with respect to the scatterers and reflecting surfaces such as ceiling, side and opposite walls, etc. In this physical model, each ray is determined by its parameters defined by its delay, azimuth-co-elevation angle of arrival, and azimuth-co-elevation angle of departure. The complex amplitude of each ray is computed with electromagnetic formulations for free space loss and loss due to the interaction with the scatterers in the environment. The interaction loss depends on the interaction, wave-front and geometrical properties of impinging rays and physical properties of reflecting surfaces. Different coefficients can be used to capture the interaction losses that depend on the transmit waveform type: plane wave, cylindrical wave or spherical wave. The most commonly used reflection coefficient is the Fresnel plane wave reflection coefficient, which is valid for flat surfaces and is function of the incidence angle and electrical properties of the reflecting surface. The RF propagation characteristics depend on how the multi-ray components interact with each other based on their phases and amplitudes constructively or destructively to create different fading profiles. The multi-domain RF characteristics depend on the dispersion pattern in their corresponding domain such as delay, direction and Doppler.

The received signal is obtained as a sum of multi-ray components as linear superposition of N individual rays, and it can be represented as follows:

$$h(t) = \frac{\lambda}{4\pi\eta_{os}}\sqrt{G_{tx}(\varphi_{los},\vartheta_{los})G_{rx}(\phi_{os},\theta_{los})} + \sum_{n=1}^{n}\frac{\lambda}{4\pi r_n}\sqrt{G_{tx}(\varphi_n,\vartheta_n)G_{rx}(\phi_n,\theta_n)}\prod_{p=1}^{P_n}\Gamma_p\, e^{jk\,d_{n,p}}$$
$$\times\,\delta(t-\tau_{los})e^{-jk(r_{los}-V\cdot\Psi_{los}t)} \qquad\qquad \times\,\delta(t-\tau_n)e^{jk(V\cdot\Psi_n)t}$$
,

where k is the wave number expressed as $k = \frac{2\pi}{\lambda}$, λ is the wavelength of operating frequency, Γ_p denotes the Fresnel reflection coefficient for the p-th wave-interface intersection, $G_{tx}(\varphi_n,\vartheta_n)$, $G_{rx}(\phi_n,\theta_n)$ are the transmitter and receiver antenna gain, respectively, V is the velocity (speed and direction) of the mobile terminal, which is assumed as the receiver in this notation, and defined by $V = v_x\vec{x} + v_y\vec{y} + v_z\vec{z}$ and Ψ_n stands for the arrival direction vector defined for ray n as
$$\Psi_n = \cos(\phi_n)\sin(\theta_n)\,\vec{x} + \sin(\phi_n)\sin(\theta_n)\,\vec{y} + \cos(\theta_n)\vec{z}$$

where ϕ_n and θ_n are the horizontal and co-elevation arrival angles of ray n (or LOS ray when subscript is los) relative to the x-axis and z-axis, respectively, and r_n is the path length of ray n, $d_{n,p}$ denotes the distance traversed by the specular wave between the (p - 1) and p-th boundary intersections, η_{los} is the length of LOS path and r_n is the specular reflection path length.

3. Reconfigurable Antenna (RA)

It can be used in a way to change radio channel characteristics in favor of enhancing performance of wireless communications system. The RA, per control, can change its field pattern, pointing direction, operating frequency and polarization. The key part in gaining benefit from RA is based on understanding the interplay between superposition of complex signals of multipath components of radio channel, three dimensional antenna patterns, and velocity of mobile terminal. This interplay would affect channel characteristics in delay, direction and Doppler domains, which will have impact on their corresponding correlation parameters such as coherence spectra (i.e., frequency correlation), spatial correlation and coherence time, respectively. The RF agility of RA can be used to change some of these channel correlation properties such as the coherence time and power weighted multipath dispersion metrics. In this work, reconfiguring antenna pattern of a mobile device is simulated in terms of rotation angle of the antenna, which changes antenna pattern for vertical polarization that can be written as [6]

$$G_v(\theta,\varphi) = 1.64\,(\cos\theta\,\cos\varphi\,\sin\alpha \;-\; \sin\theta\,\cos\alpha)^2\,\frac{\cos^2(\pi\zeta/2)}{(1-\zeta^2)^2}$$

and its antenna gain pattern for horizontal polarization is

$$G_h(\theta,\varphi) = 1.64\,\sin^2\varphi\,\sin^2\alpha\,\frac{\cos(\pi\zeta/2)^2}{(1-\zeta^2)^2}$$

where $\zeta = \sin\theta\,\cos\varphi\,\sin\alpha + \cos\varphi\,\cos\alpha$ and the angle α is the rotation angle of the antenna element from z-axis in the vertical zx-plane, φ is the azimuth angle relative to x-axis, θ is the elevation angle relative to z-axis, the coefficient 1.64 corresponds to the directivity of the half-wavelength dipole antenna.

4. Amount of Fading

The selected measure of severity of fading in this work is the amount of fading (AF), which can be computed using the first and second central moments of SNRs at diversity output. The AF is defined in [7-9] as

$$AF = \frac{var\{\alpha^2\}}{E[\alpha^2]^2}$$

where α is the instantaneous fading amplitude of a complex fading channel, $E\{\bullet\}$ and $var\{\bullet\}$ are the statistical mean and variance, respectively. To quantify the probability distribution of fading, it is mentioned in [7,8] that for Nakagami-m fading distribution of α, the amount of fading (AF), $AF = 1/m$, whose range is [0, 2]. When $m = \infty$, AF=0, which corresponds to the situation of "no fading". When $m = 1$, the AF = 1, which corresponds to Rayleigh fading, and when $m = 0.5$, AF = 2, which corresponds to the one-sided Gaussian distribution and the severest fading assumed by the Nakagami-m fading channel. Hence, when AF<1, the fading severity of the radio channel is considered less than that of Rayleigh channel and the otherwise AF>1. It is expected that the communication system performance degradation increases with AF. It is worth to mention that the fading parameter in line of sight (LOS) propagation, K factor, of Rician fading distribution is a function of the AF. The K factor is defined as the ratio of power received via LOS propagation to the power of non-LOS paths. This relationship between AF and K parameter is presented in [10] as follows

$$K = \frac{\sqrt{1-AF}}{1-\sqrt{1-AF}}$$

Different forms of this relationships are also given in [11-13]. So, the estimation of AF can be used to indicate K parameter of Rician fading channel as well as m parameter of Nakagami-m fading channel since AF=1/m. The AF formulation in [14] is presented in closed form expression for identically distributed spatially correlated Nakagami-m fading fading channel at output of at the output of a space–time block-coded multiple-input–multiple-output (MIMO) diversity system. A closed-form expression for high order amount of fading for Nakgami-m fading channel is presented in [15], which was defined originally in [16].

5. Numerical Results

The presented results are obtained for three different indoor environments, where the IEEE802.11ac system may operate. These indoor environments are 1) lecture hall with dimensions, height (H) =4 m, width (W) = 8 m, length (L) = 10 m, 2) corridor with dimensions: H = 4 m, W = 2 m, and L = 30 m, and 3) banquet hall with dimensions: H = 10m, W =15 m, and L = 50. The simulations are set-up for a Wi-Fi antenna access point (AP) at different eights (¼ H, ½ H, ¾ H, and H) and client station antenna height is 1.7 m. It is assumed that AP is the transmitter and client station is the receiver. The receiver speed is 3 km/hr, which is defined as the pedestrian speed in 3GPP standard [17]. Number of source images per reflecting surface is set to 6 that cause multiple reflection rays in addition to the line of sight component. Reflecting surfaces have relative permittivity of 5 and conductivity of 0.02. The simulated temporal range is for one second for every spatial location. The temporal sampling rate is 26,000 samples/sec. In order to investigate the effect of the environment on similar communication link setups, the three indoor environments have been tested for same route from the access point. A route starting from a horizontal distance of 2 m from AP till 9.5 m with spatial resolution of about 2.5 cm is used in the study. It is assumed that transmitter antenna is vertically polarized. Re-configurability of antenna pattern is done via controlling rotation angle. Three different rotation angles have been selected to generate three different antenna patterns for vertically polarized antennas. Figure 1 shows the three antenna patterns for vertical polarization state for three antenna states correspond to three rotation angles; 0°, -15°, and -55°. Figure 2 shows samples of fading profile for 0° rotation angle in a lecture hall for four different heights of access points. The severity of fading levels in fading profiles are computed with AF. The AF values are shown on Figure 2 as follows: 1) for AP height at ¼ H, the AF = 0.34, 2) for AP height ½ H, AF = 0.22, 3) for AP height ¾ H, the AF = 0.6, and

4) for AP height H, the AF = 0.89. Hence, it can be observed that for this particular sample, the fading profile of AP placed at ceiling has highest fading level and lowest fading level is observed when the AP is placed at height half the ceiling height. From this analysis, we see that we can map fading profiles as shown in Figure 2 from time series presentation to indices that indicate the fading severity levels. In order to get the feeling of AF numbers with severity of different fading channel models, Figure 3 shows cumulative distribution function (CDF) of envelopes of Rayleigh, double Rayleigh and Rician fading channels. The AF of Rayleigh fading is one while that of the double Rayleigh fading channel is three. Rician fading channels for every k parameter, it becomes Rayleigh as can be seen in Figure 3 for K = - 10 dB, while for AF decreases with increasing of K factor.

In order to get statistically significant results, we need to compute the AF for large number of positions within the indoor environment for different heights of AP and different rotation angles. The AF values are presented in terms of CDF. Figure 4 shows CDF of AF for antenna rotation angles in the three different indoor environments. Table I summarizes the 90th percentile of CDF curve for rotation angle of 0°. It can be seen that the curve of least AF values correspond to AP of height ½ H, ½ the ceiling height, for lecture hall and corridor environments but not for banquet hall. The height of AP that correspond to lowest AF in banquet hall is the ¼ H. However, the difference in the 90% percentile values of ¼ H and ½ H for banquet hall might be considered small. Very similar observed results can be seen for clients' antenna rotation angles at -15° as depicted on Figure 5 and presented in Table II. For client antenna's rotation angle at -55°, the lowest AF curve is observed to correspond to AP height on ceiling for lecture hall and corridor indoor environments, while for banquet hall, it is still the AP height of ¼ H has the lowest AF curve. This clearly shown on Figure 6. Table III presents the 90th percentile of AF in the three indoor environments for antenna's rotation angle of -55°. Lecture hall and corridor shows quite similar results of AF at this percentile and the value is close to that of Rayleigh fading channel. The significant corresponding values are clearly observed for banquet hall for the three AP antenna heights. It should be remembered that the curve of lowest AF corresponds to channels that have highest performance of wireless communication system. The other parameter that describes the stationarity of the fading process is the coherence time of the radio channel and how it is related to AF since AF is computed for particular time interval of channel series. Coherence time is a measure of similar behavior of radio channel over a particular period. We compute it here from autocorrelation of time series of channel envelopes at what time the normalized autocorrelation coefficient starts to be below 0.5. We extracted the coherence times for all channel positions, where we already computed the AF values. Then, we calculated the cross-correlation level between the AF and coherence time. Table IV presents correlation levels for the three indoor environments and the four AP antenna heights for the case, when antenna's rotation angle is 0°. It can be seen that the highest absolute correlation levels correspond to the case when the AP antenna height is ½ the ceiling height. The values indicate that the as the AF gets lower the coherence time gets longer, which is expected since the low AF means the that radio channel is less sever fading and stability condition is what makes the coherence time longer. The absolute correlation values for banquet hall are largest for AP heights of ¼ H and at ceiling. Their correlation values are quite similar but the curve of AF of AP height of ¼ H is smallest. Table V presents correlation levels between channel coherence time and AF for the three indoor environments when client antenna's rotation angle is -15°. It shows very similar results as in the case of rotation angle of 0°. Table VI presents the corresponding correlation values for rotation angle of -55°. The larges absolute correlation values is observed for the case of AP antenna is placed on ceiling of lecture hall and corridor. While for banquet hall, the largest absolute correlation value is for the case of AP antenna height is at ¼ H.

6. CONCLUSIONS

This work showed that impact of client antenna re-configurability in terms of its rotation angle and impact of antenna height of AP on level of AF in three different indoor environments of different dispersion characteristics. The results show that for placing antenna of AP at height ½ of the ceiling height in lecture hall and corridor leads to least AF, which may lead to higher performance of wireless communication system. This is true when rotation angle is close to vertical polarization, i.e., 0° and -15°, but the effect of AP height on AF is quite small when rotation angle is -55°. For banquet hall, the results show that placing AP antenna height at ¼ H shows lowest ranges of AF at antenna's rotation angles of 0° and -15° and when AP is on the ceiling results in smallest values of AF for rotation angle of -55°.

ACKNOWLEDGEMENTS

This publication was made possible by NPRP grants #: NPRP 5-653-2-268 from the Qatar National Research Fund (a member of Qatar Foundation). The statements made herein are solely the responsibility of the authors.

REFERENCES

[1]   Official IEEE 802.11 Working Group Project Timelines-2013-05-18, Online: http://grouper.ieee.org/groups/802/11/Reports/802.11_Timelines.htm.

[2]   Greg Breit, et al, "TGac Channel Model Addendum", Institute of Electronic and Electrical Engineers, IEEE802.11-09/0308r12, March 18, 2010. Online: Available: https://mentor.ieee.org/802.11/dcn/09/11-09-0308-12-00ac-tgacchannel-model-addendum-document.doc.

[3]   Breit, G. et al., "TGac Channel Model Addendum Supporting Material," Doc. IEEE802.11-09/0569r0. Erceg,

[4]   V. et al. "TGn Channel Models." Doc. IEEE802.11-03

[5]   W. Q. Malik, C. J. Stevens, and D. J. Edwards, "Spatio-temporal ultrawideband indoor propagation modelling by reduced complexity geometric optics," *IET Commun.*, vol. 1, no. 4, pp. 751-759, 2007.

[6]   T. Taga "Analysis for Mean Effective Gain of Mobile Antennas in Land Mobile Radio Environments," *IEEE Trans. Veh. Technol.* vol. 39, no. 2, pp. 117 – 131, 1990.

[7]   U. Charash, "Reception through Nakagami multipath channels with random delays," IEEE Trans. on Commun., vol. 27, no. 4, pp. 657–670, Apr. 1979.

[8]   M. Simon and M. Alouini, Digital Communications over Fading Channels: A Unified Approach to Performance, Analysis, John Wiley & Sons, Inc. 2000.

[9]   H. El-Sallabi, K. Qaraqe and Erchin Serpedin, "Some Insights on the Amount of Fading in Radio Channels" in Proc. PIERS 2013, Stockholm, Sweden, August, 2013.

[10]   A. Abdi, A., C. Tepedelenlioglu, M. Kaveh, and *G. Giannakis,* "On the estimation of the K parameter for the Rice fading distribution", IEEE Communications Letters, Volume 5, Number 3, March 2001, Pages 92–94.

[11]   P. K. Rastogi and O. Holt, "On detecting reflections in presence of scattering from amplitude statistics with application to D region partial reflections," *Radio Sci.*, vol. 16, pp. 1431–1443, 1981.

[12]   L. J. Greenstein, D. G. Michelson, and V. Erceg, "Moment-method estimation of the Ricean K-factor," *IEEE Commun. Lett.*, vol. 3, pp. 175–176, 1999.

[13]   P. D. Shaft, "On the relationship between scintillation index and Rician fading," *IEEE Trans. Commun.*, vol. 22, pp. 731–732, 1974.

[14]   B. Holter and G. E. Øien "On the Amount of Fading in MIMO Diversity Systems," *IEEE Trans. On Wireless Commun.*, vol. 4, no. 5, Sep. 2005, pp. 2498 – 2507.

[15]   A. Hyadi, M. Benjillali, M.-S. Alouini, and D. B. da Costa, "Performance Analysis of Underlay Cognitive Multihop Regenerative Relaying Systems with Multiple Primary Receivers" *IEEE Trans. Wireless Commun.*, vol. 12, pp. 6418–6429, 2012.

[16]   F. Yilmaz and M.-S. Alouini, "Novel asymptotic results on the high order statistics of the channel capacity over generalized fading channels," in *Proc. 2012 International Workshop on Signal Processing Advances in Wirless Communications*, pp. 389–393.

[17]   http://www.3gpp.org/

Table I. The 90th percentile of AF for antenna's rotation angle of $0°$.

	¼ H	½ H	¾ H	H
Lecture Hall	1.14	0.81	1.21	1.14
Corridor	0.89	0.75	0.96	1.10
Banquet Hall	0.71	0.84	1.03	1.11

Table II. The 90th percentile of AF for antenna's rotation angle of $-15°$.

	¼ H	½ H	¾ H	H
Lecture Hall	1.17	0.84	1.14	1.02
Corridor	0.92	0.79	0.89	0.98
Banquet Hall	0.66	0.69	0.79	0.83

Table III. The 90th percentile of AF for antenna's rotation angle of $-55°$.

	¼ H	½ H	¾ H	H
Lecture Hall	1.03	1.03	1.06	0.99
Corridor	0.95	1.10	0.93	0.99
Banquet Hall	0.74	0.56	0.44	0.42

Table IV. Correlation value between channel coherence time and AF for antenna's rotation angles of $0°$.

	¼ H	½ H	¾ H	H
Lecture Hall	-0.55	-0.62	-0.5	-0.33
Corridor	-0.68	-0.79	-0.69	-0.65
Banquet Hall	-0.57	-0.34	-0.50	-0.60

Table V. Correlation value between channel coherence time and AF for antenna's rotation angles of $-15°$.

	¼ H	½ H	¾ H	H
Lecture Hall	-0.57	-0.61	-0.46	-0.42
Corridor	-0.64	-0.78	-0.69	-0.70
Banquet Hall	-0.61	-0.43	-0.51	-0.61

Table VI. Correlation value between channel coherence time and AF for antenna's rotation angles of $-55°$.

	¼ H	½ H	¾ H	H
Lecture Hall	-0.48	-0.54	-0.60	-0.81
Corridor	-0.40	-0.75	-0.46	-0.73
Banquet Hall	-0.69	-0.65	-0.63	-0.20

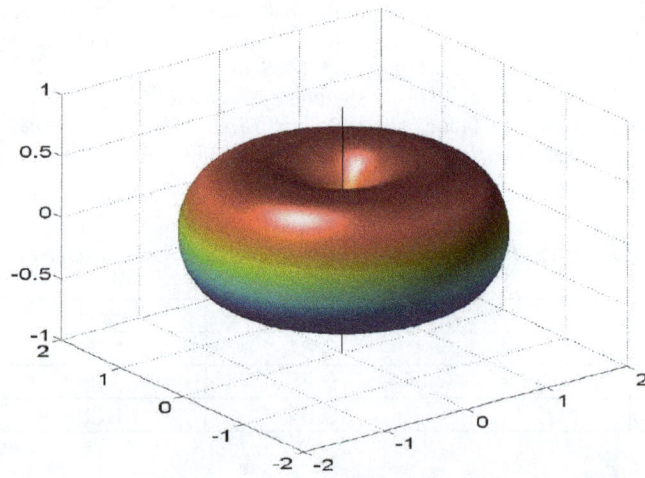

a.   Rotation angle = 0°

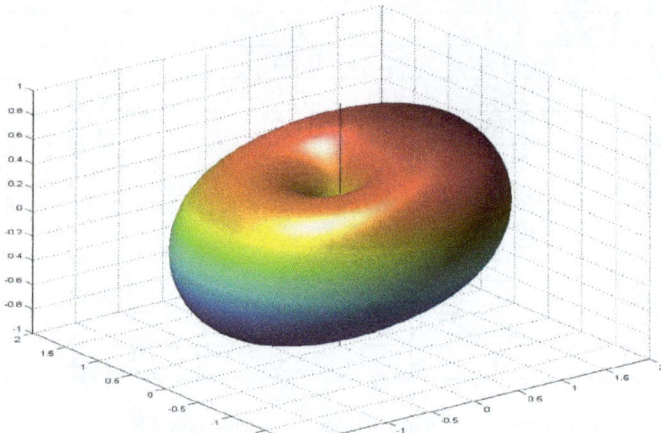

b.   Rotation angle = -15°

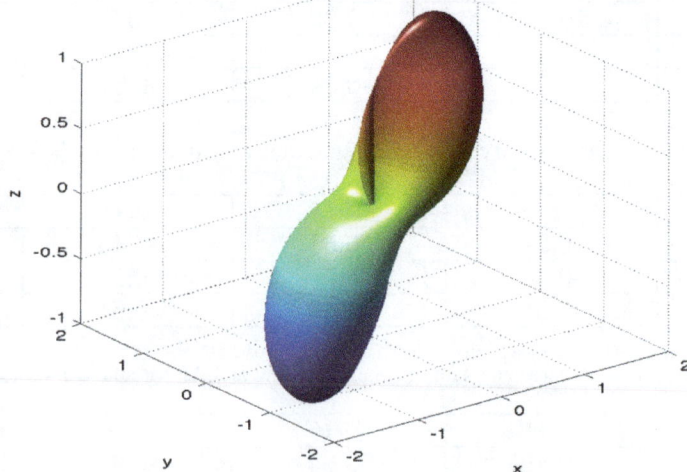

a.   Rotation angle = -55°

Figure 1. Antenna pattern for vertical polarization at different antenna's rotation angles

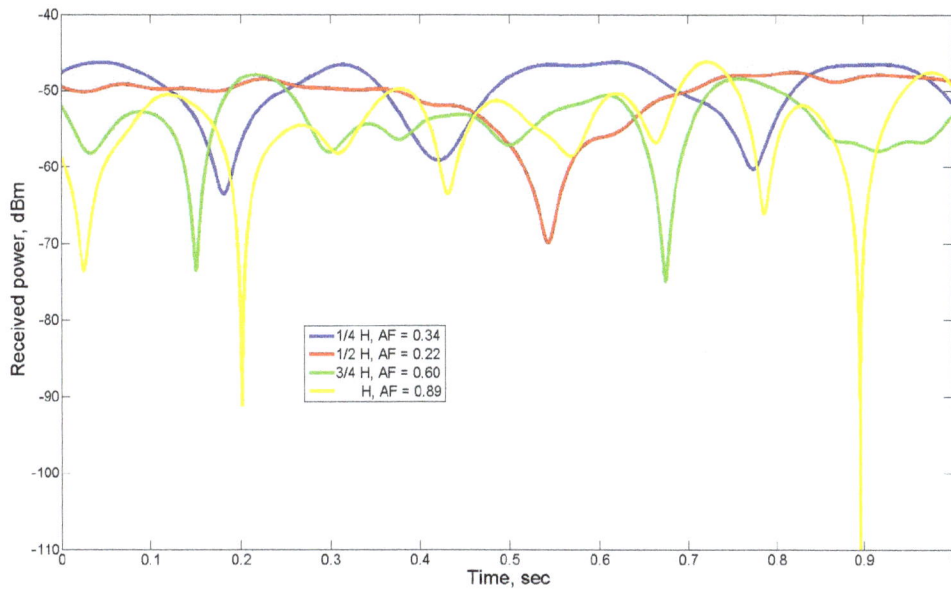

. Figure 2. Samples of fading profiles and corresponding AF values

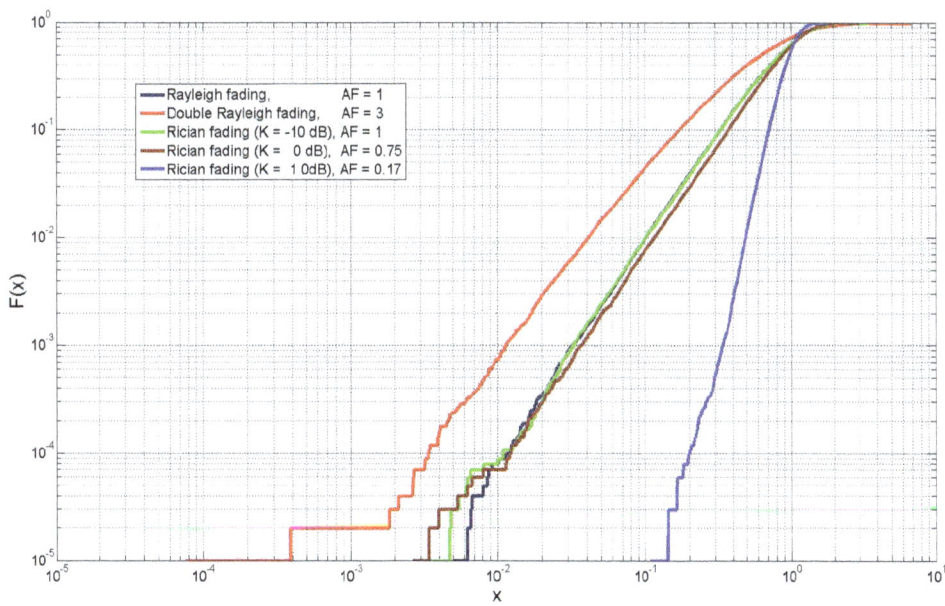

Figure 3. CDF of different common fading channels with their corresponding AF values.

a.   Lecture Hall

b.   Corridor

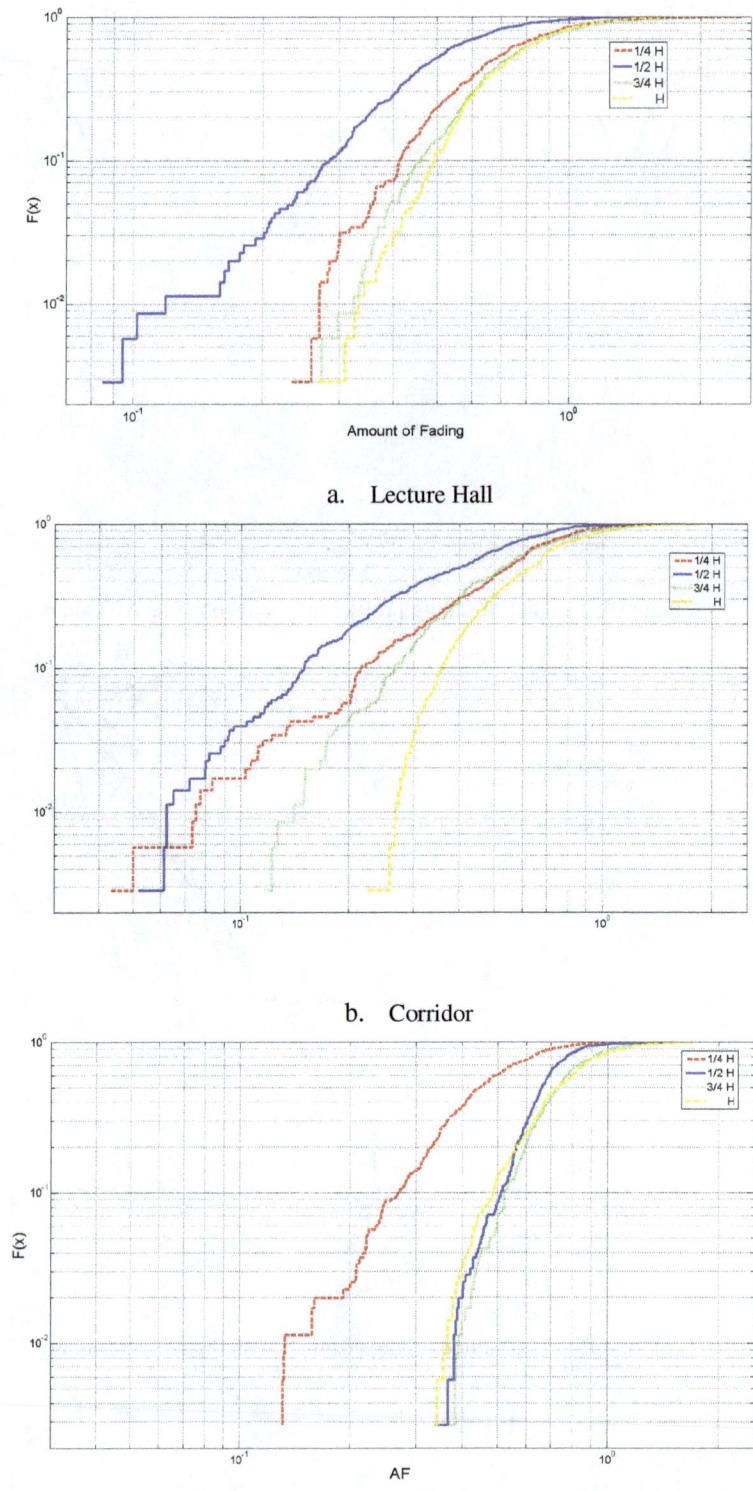

c.   Banquet Hall
Figure 4. Impact of antenna height of AP on AF in three different indoor environment for 0o rotation angle of antenna at mobile station

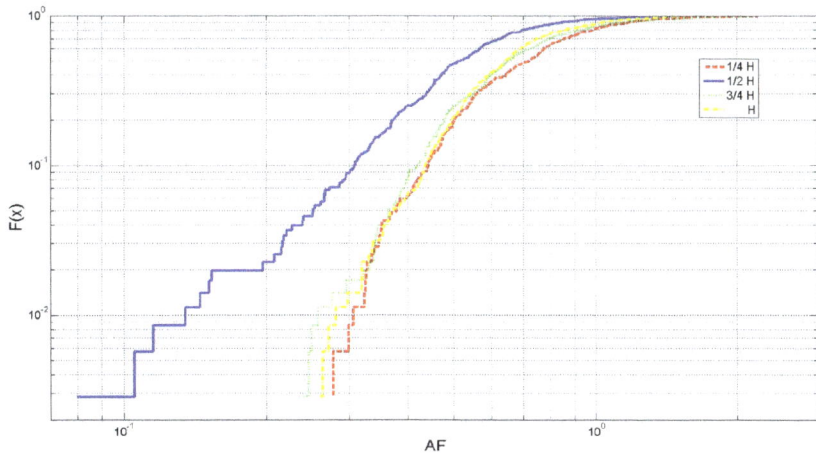

a. Lecture Hall

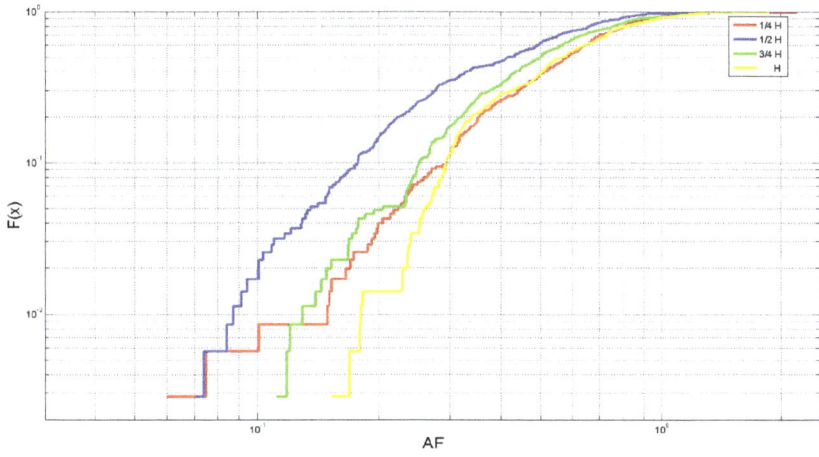

b. Corridor

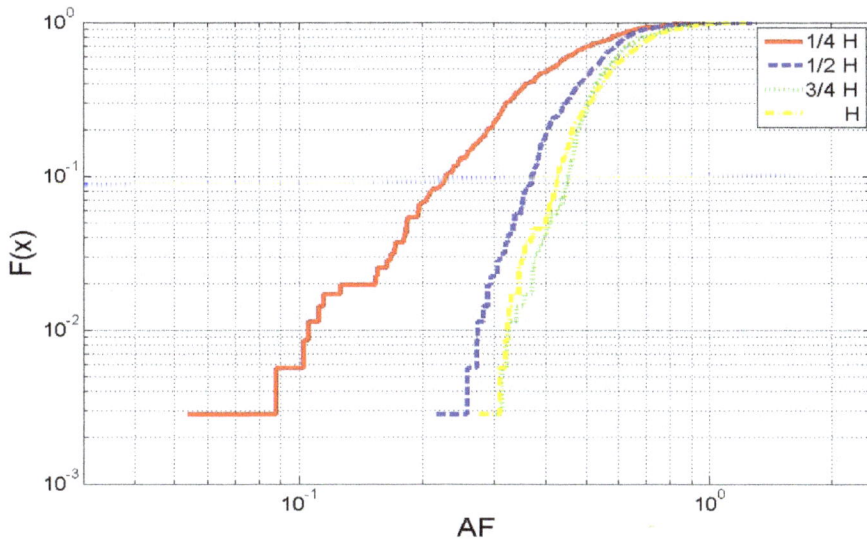

c. Banquet Hall

Figure 5. Impact of antenna height on AF in three different indoor environment for -15° rotation angle of antenna at mobile station.

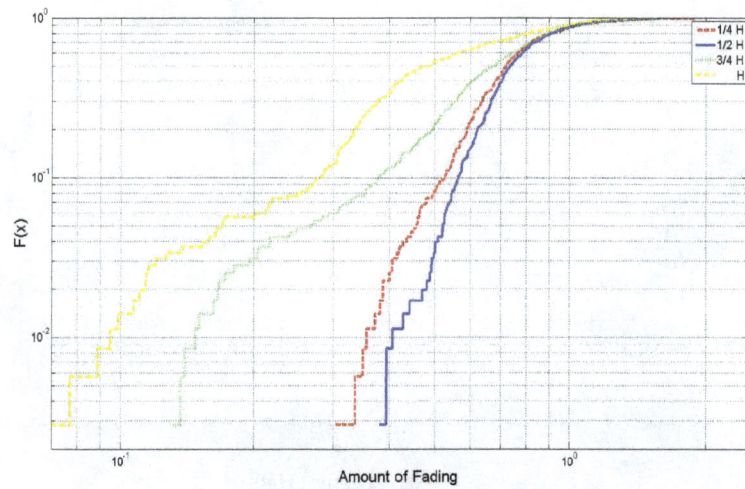

a.   Lecture Hall

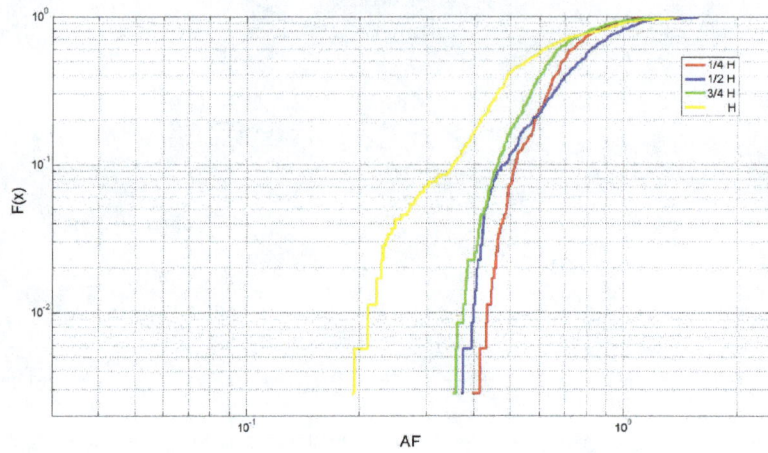

b.   Corridor

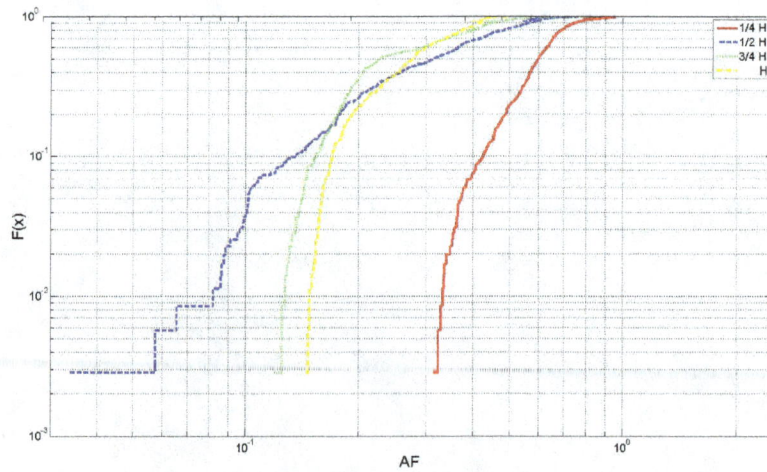

c.   Banquet Hall

Figure 6. Impact of antenna height on AF in three different indoor environment for -55° rotation angle of antenna at mobile station.

15

DESIGN AND ANALYSIS OF MIMO SYSTEM FOR UWB COMMUNICATION

Mihir N. Mohanty[1], Monalisa Bhol[2], Laxmi Prasad Mishra[3], Sanjat Kumar Mishra[4]

1IITER, Siksha 'O' Anusandhan University, Bhubaneswar, Odisha, 751030, India

Seemanta Engineering College, Jharpokharia, Mayurbhanj, Odisha, India

ABSTRACT

Multiple transmit and receive antennas are used MIMO system. The system creates parallel MIMO subchannels to transmit independent streams of data under the appropriate channel conditions. Similarly, Ultrawideband (UWB) communication has attracted great interest for various applications in recent days. Spatially multiplexed (SM) multiple-input multiple-output (MIMO) systems gains the spectral efficiency as well as high data rates without consuming additional power, bandwidth or time slots. In this paper, we extend the concept of MIMO to UWB systems. The correlated channel for such purpose is considered and the performance has been analyzed for spatial multiplexing SM-UWB-MIMO system which is required for estimation. The system performance substantially degrades in the presence of high values of spatial correlation. To avoid the degradation of such system, it has been designed for virtual UWB-MIMO Time Reversal (TR) system, so that it is not affected by the transmit correlation. Another novel method to reduce the effect of correlation has been chosen by taking the Eigen value of the channel matrix for the computation of the system performance. The result shows its performance.

KEYWORDS

MIMO, UWB, BER, Channel Capacity, Spatial Correlation, TR, Eigen Value.

1 INTRODUCTION

Wireless propagation channels have been investigated for more than two decades and a large number of channel models are designed by many researchers. Ultrawideband (UWB) communication system has become most promising for high data rate as well as short-range communication systems. Therefore, it has attracted great interests from both academic and industrial aspects recently. Because of the restrictions on the transmit power, UWB communications are best suited for short-range communications [1].

Increasing demand for higher wireless system capacity has catalyzed several transmission techniques, among which multiple-input/multiple-output (MIMO) technology is popular one. Extending MIMO technology to the UWB regime, a large gain in the channel capacity, robustness and coverage radius is noticed in UWB indoor communications systems [2]. These systems are equipped with multiple antennas, at both the transmitter and receiver in order to improve communication performance, in contrast to conventional communication systems with only one antenna on the transmitter and one antenna on the receiver.

MIMO technologies overcome the deficiencies of the traditional methods through the use of spatial diversity. Data can be transmitted over M transmit antennas to N receive antennas supported by the receiver terminal. Such systems are used in wireless communication for enhancement of capacity and bit error rate (BER). It offers significant increases in data throughput and link range without additional bandwidth or transmit power. These characteristics are essential for the coming generation of Telecommunications systems. Rayleigh fading has been considered as the propagation channel for verification. Diversity gain and spatial multiplexing (SM) are the two main advantages of MIMO systems that are used to study the effect of increase in bit rate with increasing the number of transmitter and receiver antennas. In MIMO system, we primarily need to take into account the spatial correlation. The effect of spatial correlation has to be minimized to obtain better system performance. In [3], the time-reversed channel impulse response (CIR) is implemented as a filter at the transmitter side. It is well known that the MIMO-TR-UWB system can achieve transmit diversity, but it suffers from both transmit and receive antenna correlations. The single-input multiple-output TR-UWB (SIMO-TR-UWB) or virtual MIMO-TR-UWB does not face the transmit antenna correlation because it has only one transmit antenna.

2. RELATED LITERATURE

An overview of reported measurements and modelling of the UWB indoor wireless channel is presented in [4]. Different UWB channel sounding techniques are discussed and approaches for the modelling of the UWB channel are reviewed.

A considerable work has been performed in [5-7] to characterize communication channels for general wireless applications. As MIMO systems operate at an unprecedented level of complexity to exploit the channel space-time resources, a new level of understanding of the channel space-time characteristics is required to assess the potential performance of practical multi-antenna links.

Empirical investigation of spatial correlation in UWB indoor channels has been presented in [8]. It was observed that the coherence distance falls with channel bandwidth in end-fire arrays but not in broadside arrays. The complex correlation decays less rapidly with distance in broadside arrays than in end fire arrays, especially under line-of-sight. Strong dependence of spatial correlation and coherence distance on the channel centre frequency was observed.

Spatial multiplexing single-input–multiple-output (SM-SIMO) UWB communication system using the TR technique has been proposed. The system with only one transmit antenna, using a spatial multiplexing scheme, can transmit several independent data streams to achieve a very high data rate. TR can mitigate not only the ISI but the MSI caused by multiplexing several data streams simultaneously as well [9]. Antenna selection scheme for MIMO UWB communication system with TR is investigated in [10].

Time reversal technique has advantage in highly scattering environments to achieve signal focusing through transmitter-side processing that enables the use of simple receivers. The authors have also demonstrated UWB time reversal system architecture taking into account some practical constraints [11].

3. METHODOLOGY

A tractable correlated MIMO UWB channel model is essential when developing multiple-antenna UWB systems in order to accurately predict their performance. The system is designed based on Alamouti code. Figure 1 show for a 2X2 antenna system. In Alamouti encoding scheme, during any given transmission period two signals are transmitted simultaneously from two transmit antennas.

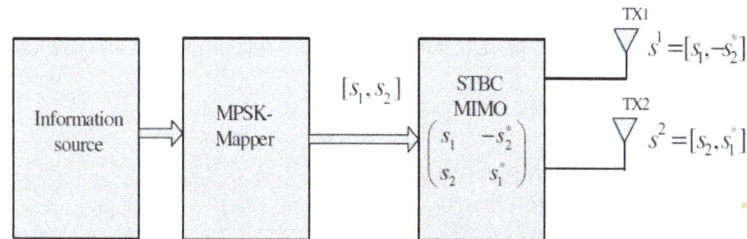

Fig. 1. Encoder for Alamouti schemes

At time t, antenna 1 transmits s1, and simultaneously, antenna 2 transmits s0. At time t + T, where T is the symbol duration, signal transmission is switched, with –s0* transmitted by antenna 1 and s1* simultaneously, transmitted by antenna 2. We present the measurements of a MIMO system under line-of-sight conditions.

3.1 Spatial Correlation

Though the space-time focusing feature is one of the benefits, spatial multiplexing has a major role in MIMO systems. Without the expansion of bandwidth, high data rate can be achieved by using spatial multiplexing scheme with multiple transmit and receive antennas. The spatial correlation in the multipath channel is a critical factor in the performance of a MIMO system and is evaluated. It is mainly caused by inadequate antenna spacing in both transmitting and receiving side. It causes correlation between the received signals, which degrades the signal quality, capacity and bit error rate (BER) performance. Capacity increases and BER performance also increases as signal correlation decreases. The fading correlation between the array elements should be sufficiently low for a MIMO system to offer any performance enhancement.

The MIMO channel matrix is assumed to be independent of each other. Other assumptions that can be made for such model analysis is as follows:

The correlation among the receive antennas is independent of the correlation between the transmit antennas.

The effect of antenna coupling is neglected, and we focus only on the spatial correlation.
The transmit and receive correlation matrices are fixed.

The correlation has been included to the MIMO UWB channel model by introducing fixed transmit and receive correlation matrices following the well-known Kronecker model [12-14],

$$H = R_{rx}^{1/2} H_w R_{tx}^{1/2} \tag{1}$$

where, H_w = channel matrix of independent channel realization.

$R_{tx}^{1/2}$ = transmit correlation matrix with dimension $M \times M$.

$R_{rx}^{1/2}$ = receive correlation matrix with dimension $N \times N$.

H = correlated channel

3.2 Methods for Computing Spatial Correlation

1. Gathering a large amount of data in the target propagation environment. For that it becomes necessary to estimate a large number of correlation coefficients in an M x N MIMO system. Suppose, there are MN spatial sub-channels, and correlating each pair of them would give rise to $(MN)^2$ correlation values. A disadvantage of this approach, beside the fact that it may be very time-consuming, is that it may be necessary to estimate a large number of correlation coefficients.

2. Using fixed correlation matrices as:

    i.   Transmit correlation matrix
    ii.   Receive correlation matrix

The transmit correlation matrix is given as:

$$R_{tx} = \begin{bmatrix} 1 & \rho_{Tx} & \rho_{Tx}^2 \cdots & \rho_{Tx}^{M-1} \\ \rho_{Tx} & 1 & \rho_{Tx} \cdots & \rho_{Tx}^{M-2} \\ \vdots & \vdots & \ddots & \vdots \\ \rho_{Tx}^{M-1} & \rho_{Tx}^{M-2} & \rho_{Tx}^{M-3} \cdots & 1 \end{bmatrix} \tag{2}$$

Similarly, receive correlation matrix is given as:

$$R_{rx} = \begin{bmatrix} 1 & \rho_{Rx} & \rho_{Rx}^2 \cdots & \rho_{Rx}^{N-1} \\ \rho_{Rx} & 1 & \rho_{Rx} \cdots & \rho_{Rx}^{N-2} \\ \vdots & \vdots & \ddots & \vdots \\ \rho_{Rx}^{N-1} & \rho_{Rx}^{N-2} & \rho_{Rx}^{N-3} \cdots & 1 \end{bmatrix} \tag{3}$$

The advantage of the fixed correlation model is its simplicity and its immediate application to the existing IEEE 802.15.3a standard.

The fixed correlation matrices R_{tx} and R_{rx} appropriate for a particular environment can be determined by selecting the numerical values of ρ_{Tx} (transmit correlation coefficient) and ρ_{Rx} (receive correlation coefficient), such that a close match is obtained to the BER results achieved when conducting the system simulation using the measured indoor channel. The correlation coefficient values ranges from 0 to 1. We present BER results for various MIMO UWB systems for various values of the channel correlation coefficient.

3.3 Proposed Method for Reduction of Correlation Effect

    1.   Virtual MIMO-UWB-Time Reversal Technique System

The time reversal (TR) technique, which is originated from under-water acoustics and ultrasonic, now has been used in many applications such as localization, imaging and green wireless communications. TR also has shown its potential in dealing with the ISI problems in UWB. In a TR system, the time-reversed channel response (CIR) is implemented as a filter at the transmitter side [15].

With the help of the TR pre-filter, the system with only one transmit antenna can deliver several independent data streams at the same time. We have taken the spatial correlation into account in, where a constant spatial correlation model for MIMO UWB has been applied. The performance of system over correlated line of- sight (LOS) channels is investigated with the same correlation model. This scenario is referred to as channel model1 (CM1) in the IEEE 802.15.3a standard. The BER results on the adopted correlated channel model with an appropriate value of correlation coefficient are shown closely matching with those on the measured indoor channel.

It is well known that the MIMO-TR-UWB system can achieve transmit diversity, but it suffers from penalty caused by both transmit and receive antenna correlations. Meanwhile, the single-input multiple-output TR-UWB (SIMO-TR-UWB) or virtual MIMO-TR-UWB does not face the transmit antenna correlation because it has only one transmit antenna. It is shown the virtual MIMO outperforms the true MIMO system in terms of the BER performance. The channel impulse response (CIR) between the transmit antenna j and the receive antenna i is, [15],

$$h_{i,j}(t) = \alpha^{i,j} \delta(t - \tau^{i,j}) \qquad (4)$$

where, α is the amplitude

τ is the delay and the value is considered for it as IEEE 802.15.3a standard from the Table-1

$$H_{i,j}(t) = \begin{pmatrix} h_{1,1}(t) & h_{1,2}(t) & \cdots & h_{1,M}(t) \\ h_{2,1}(t) & h_{2,2}(t) & \cdots & h_{2,M}(t) \\ \vdots & \vdots & \ddots & \vdots \\ h_{N,1}(t) & h_{N,2}(t) & \cdots & h_{N,M}(t) \end{pmatrix} \qquad (5)$$

The TR pre-filter matrix of the MIMO system is given by:

$$\overline{H_{i,j}}(t) = \begin{pmatrix} \overline{h_{1,1}}(-t) & \overline{h_{2,1}}(-t) & \cdots & \overline{h_{N,1}}(-t) \\ \overline{h_{1,2}}(-t) & \overline{h_{2,2}}(-t) & \cdots & \overline{h_{N,2}}(-t) \\ \vdots & \vdots & \ddots & \vdots \\ \overline{h_{1,M}}(-t) & \overline{h_{2,M}}(-t) & \cdots & \overline{h_{N,M}}(-t) \end{pmatrix} \qquad (6)$$

where, $\qquad \overline{h}(t) = h(t) \otimes h(-t) \qquad (7)$

The matrix of the equivalent channel is a square matrix with the entries in the main diagonal being the summation of the autocorrelation of the original CIRs and other entries being the summation   of the cross-correlation of the original CIRs between the transmit and receive antennas.

The TR matrix $\overline{H_{i,j}}(t)$ is used instead of $H_W(t)$ in equation (1), to calculate the BER performance of the MIMO UWB system. The following parameters have been considered for evaluation of UWB channel model. It has been tested according to the IEEE standard 802.15.3a UWB model.

Table 1. Channel model Parameter of IEEE 802.15.3a standard.

Parameters	Specific Values considered
Channel Model	CM1 for Line of Sight communication
Frequency	3 GHz
Channel	Rayleigh fading channel
Modulation τ	QPSK 5.05 ns

It is shown the virtual MIMO outperforms the true MIMO system in term of the BER performance. Another method to reduce the effect of correlation has been chosen by taking the Eigen value of the channel matrix for the computation of the system performance.

2.   Eigen Values of the Correlation Matrix

The sub-channel correlation, power gains of supported eigen modes, and branch power ratios are analyzed. The mutual information capacity is found to scale almost linearly with the MIMO array size, with very low variance. Eigen value of the correlation matrix is considered, that can be expressed as the relation:

$$\det(\lambda I - A) = 0 \tag{8}$$

where, λ = the eigen value of A and A = square matrix

4. RESULTS AND DISCUSSION

In Fig. 2, shows the performance of channel capacity in the wireless channel. The result is measured according to the various SNR values. The capacity of the MIMO channel has been simulated for number of transmitter and receiver antennas, such as 2 x 2, 3 x 3, 4 x 4, and 8 x 8 MIMO systems. It is observed that capacity gradually increases with the number of antennas. Fig. 3, i.e., BER versus SNR, with different number of transmitting and receiving antennas(2 x 2, 3 x 3, 4 x 4, and 8 x 8). It is seen that the BER performance increases gradually with the increase in the number of antennas.

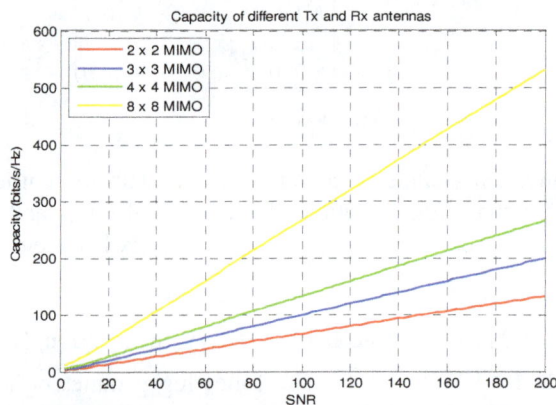

Fig. 2. Performance of capacity with respect to SNR

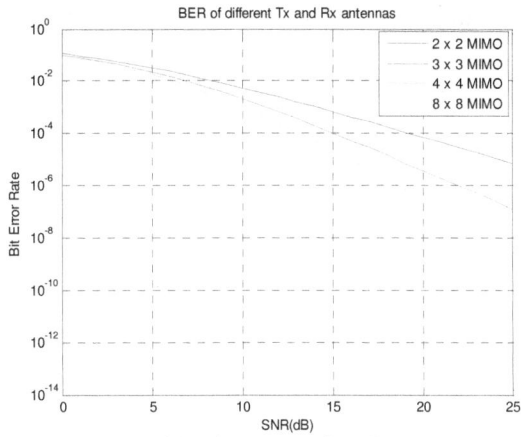

Fig. 3 Performance of BER

Fig. 4 shows the capacity results for $M = N = 2$, that is (2 x 2) for the systems operating in the CM1 with correlation coefficients ρ_{Tx} and ρ_{Rx} to be 0, 0.3 and 0.9 in the measured UWB LOS channel. The capacity for different correlation factors is tested and it has been found that the capacity decreases with increase in the correlation factors and also decreases with increase in the number of correlated antenna elements. BER results for $M = N = 2$, that is (2 x 2) for the systems operating in the CM1 with correlation coefficients ρ_{Tx} and ρ_{Rx} to be 0, 0.3 and 0.9 and without applying time reversal in the measured UWB-LOS channel is shown in Fig. 5.

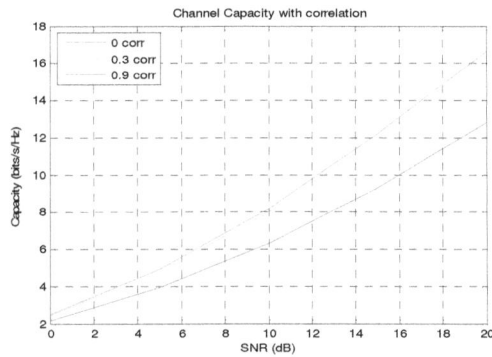

Fig.4. Capacity performance with correlation in 2 x 2 systems

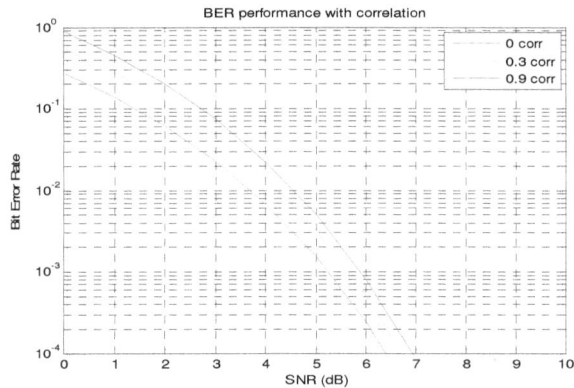

Fig. 5. BER performance with correlation

It has been observed that the BER performance decreases with increase in the value of the correlation coefficients.

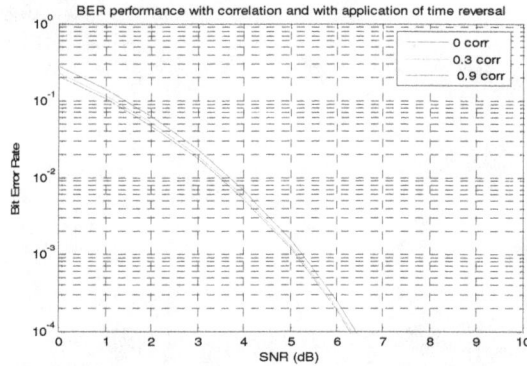

Fig. 6. BER performance with correlation and time reversal technique

The BER results for M =1, N= 2, that is (1 x 2) for the systems operating in the CM1 with correlation coefficients ρ_{Tx} and ρ_{Rx} to be 0, 0.3 and 0.9 is shown in Fig. 6, by applying virtual MIMO time reversal in the measured UWB -LOS channel. Also it has been shown that the BER performance is better than the one without time reversal.

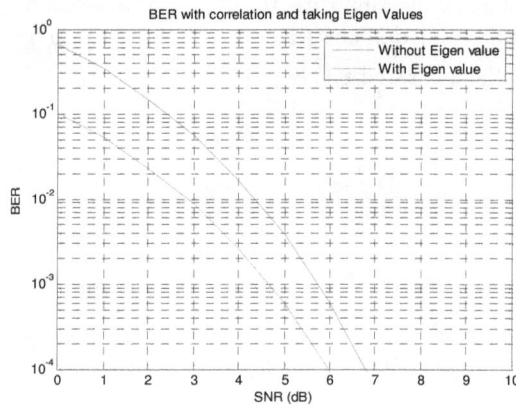

Fig. 7. BER with Eigen values

In Fig. 7, BER performance of 2 x 2 MIMO-UWB systems in the LOS indoor CM1 with correlation coefficients $\rho_{Tx} = \rho_{Rx} = 0.4$ is shown. The comparison of the BER shows its efficacy. It is considered with correlation and the Eigen value of the correlation matrix in shown. Here, it has been observed that, the BER performance is even better for eigen values of the correlation matrix if considered.

5.CONCLUSION

Multiple antenna communications technologies offer significant advantages over single antenna systems. We have proposed different methods to reduce the impact of correlation in the MIMO-UWB channel and presented a comprehensive comparative analysis with a distance independent spatial correlation model. Comparison between the BER performance of a system with correlation using virtual MIMO time reversal technique and without using time reversal technique is evaluated. BER performance of the system with eigen values and without eigen values of the correlation matrix respectively are also evaluated. Capacity result for 2 x 2 systems with different correlation coefficient values are shown as the proof. These advantages include extended range, improved reliability in fading environments and higher data throughputs.

REFERENCES

[1]   Andreas F. Molisch., Jeffrey R. Foerster.: Channel Models For Ultrawideband Personal Area Networks,Vol. 10. IEEE Wireless Communications (2003) 14-21.

[2]   Wasim Q. Malik., David J. Edwards.: Measured MIMO Capacity and Diversity Gain With Spatial and Polar Arrays in Ultrawideband Channels,Vol. 55. IEEE Transactions On   Communications (2007) 2361–2370.

[3]   T. K. Nguyen., H. Nguyen, F. Zheng., T. Kaiser.: Spatial correlation in SM-MIMO-UWB systems using a pre-equalizer and pre-Rake filter, Proceedings of the IEEE International Conference on Ultra-Wideband (ICUWB) (2010) 540–543.

[4]   Zoubir Irahhauten., Homayoun Nikookar., Gerard J. M. Janssen.: An Overview Of Ultra Wide Band Indoor Channel Measurements And Modeling, Vol. 14. IEEE Microwave And Wireless Components Letters (2004) 386-38.

[5]   Mihir Narayan Mohanty., Sikha Mishra.: Design of MCM based Wireless System using Wavelet Packet Network & its PAPR Analysis, IEEE Conference, ICCPCT-2013 (2013) 821-824.

[6]   Mihir Narayan Mohanty., Laxmi Prasad Mishra., Saumendra Kumar Mohanty.: Design of MIMO Space-Time Code for High Data Rate Wireless Communication, Vol. 3, No. 2. IJCSE, (2011) 693-696.

[7]   Michael A. Jensen., Jon W. Wallace.: MIMO Wireless Channel Modeling and Experimental Characterization, Space-Time Processing for MIMO Communications, John Wiley & Sons, Ltd (2005) 1-39.

[8]   Wasim Q. Malik.: Spatial Correlation in Ultrawideband Channels, Vol. 7. IEEE Transactions On Wireless Communications (2008) 604-610.

[9]   Hieu Nguyen., Zhao Zhao, Feng Zheng., Thomas Kaiser.: Preequalizer Design for Spatial Multiplexing SIMO-UWB TR Systems,Vol.59. IEEE Transactions On Vehicular Technology (2010) 3798-3805.

[10]   Hieu Nguyen., Feng Zheng., Thomas Kaiser.: Antenna Selection for Time Reversal MIMO UWB Systems, IEEE Transactions on Signal Processing (2009) 1-5.

[11]   Nan Guo., Brian M. Sadler., Robert C. Qiu.: Reduced-Complexity UWB Time-Reversal Techniques and Experimental Results,Vol. 6. IEEE Transactions on Wireless Communications (2007) 1-6.

[12]   T. Kaiser., F. Zheng., E. Dimitrov.: An Overview of Ultrawide-band Systems with MIMO,Vol. 97. Proceedings of the IEEE (2009) 285-312.

[13]   A. Paulraj., R. Nabar., D. Gore.: Introduction to Space-Time Wireless Communications, Cambridge University Press (2003).

[14]   S. L. Loyka.: Channel Capacity of MIMO Architecture Using the Exponential Correlation Matrix, Vol. 5. IEEE Communications Letters (2001) 369-371.

[15]   Hieu Nguyen., Van Duc Nguyen., Trung Kien Nguyen., Kiattisak Maichalernnukul., Feng Zheng., Thomas Kaiser.: On the Performance of the Time Reversal SM-MIMO-UWB System on Correlated Channels, Hindawi Publishing Corporation International Journal of Antennas and Propagation (2012) 1-8.

16

THE IMPACT OF ADAPTATION POLICIES ON CHANNEL CAPACITY OVER RAYLEIGH FADING WITH EGC DIVERSITY

Moses Ekpenyong[1], Joseph Isabona[2] and Imeh Umoren[3]

[1]Department of Computer Science, University of Uyo, Uyo
Academic Visitor, School of Informatics, University of Edinburgh, Edinburgh
v1emoses@sms.ed.ac.uk, mosesekpenyong@gmail.com
[2]Department of Basic Sciences, Benson Idahosa University, Benin City, Nigeria
josabone@yahoo.com
[3]Department of Computer Science, Akwa Ibom State University, Mkpat Enin, Nigeria
hollymeh_u@yahoo.com

ABSTRACT

In this paper, we study the basic diversity combining techniques and investigate analytically-derived closed-form expressions for the capacities per unit bandwidth for Rayleigh fading channels with equal gain combining (EGC) diversity. We consider different channel performance adaptation policies such as: power and rate adaptation with constant transmit power, channel inversion with fixed rate, and truncated channel inversion. A simulation of these schemes was carried out under ideal communication conditions and their performance compared. Results obtained show that channel inversion policies gave the highest capacity over other adaptation policies (with EGC diversity) while the constant transmit power policy yielded the lowest capacity. Furthermore, the truncated channel inversion policy outperformed other policies, while the constant power policy had the lowest capacity, compared to other policies.

KEYWORDS

Rayleigh fading, channel capacity, channel gain, multi-path fading

1. INTRODUCTION

The astronomical growth in the capacity of wireless communication services in recent times is not unconnected with the high-tech communication infrastructure which has simplified the communication process. On the other hand, this growth has generally affected the allocation of the scarce radio spectrum, due to the fact that more users now join the network. This therefore calls to question how the expected quality of service (QoS) will be sustained or guaranteed in the presence of these challenges. Some network operators attempt to confront these challenges by conserving the sharable bandwidth, but lack the required optimization skills. Thus, channel capacity is most vital in the design of wireless communication systems, as it determines the maximum attainable throughput of the system. An important concept is the coherence time [1], which is a measure of the time duration over which channel gain remains almost constant or highly correlated with a recommended correlation coefficient of above 0.5.

When a signal is transmitted over a radio channel, it is bound to experience delay, reflection, scattering and diffraction. This propagation defect also known as multi-path fading causes rapid changes in the communication environment, which introduces more complexities and uncertainties to the channel response and results in time-variation of the signal strength between the transmitter and the receiver. The urban and suburban areas where cellular phones are most often used are mostly affected by this phenomenon. A simulator offers a better understanding of this phenomenon.

Multi-path fading channel modelling traditionally focuses on physical-level dynamics such as signal strength and bit error rate (BER). The end-to-end modelling and design of systems that mitigates the effect of fading are usually more challenging than those whose sole source of performance degradation is the Additive White Gaussian Noise (AWGN). Fading mitigation is important because wireless systems are prone to fading which is also known to cause degradation in the wireless link performance, and calls for efficient fading mitigation schemes. In this paper, an extended simulation is carried out to investigate the effect of diversity level on channel capacity. We proceed as follows: a review of related works is first presented, then a discussion on the basic diversity methods, with a simulation (of the performance) of each technique. Finally, a discussion of an extensive simulation of the effect of different diversity level on the channel capacity is presented.

2. REVIEW OF RELATED WORKS

Diversity techniques are becoming well known techniques for combating the notorious effect of channel fading. This is evident in the numerous research works available in the study of channel capacity over fading channels. Initial investigation in this area (use of diversity schemes) dates back to [2], who studied the use of maximum ratio diversity combination (MRDC) technique to provide maximum capacity improvement. In [3], the theoretical spectral efficiency limits of adaptive modulation in Nakagami multi-path fading (NMF) channels are investigated. They apply the general theory developed in [4] to obtain closed-form expressions for the capacity of *Rayliegh* fading channels under different adaptive transmission and diversity combining techniques. In [5], the capacity of a single-user flat fading channel with perfect channel measurement information at the transmitter and the receiver is derived for various adaptive transmission policies. The basic concept of adaptive transmission is the real-time balancing of the link budget through adaptive variation of the transmitted power level, symbol transmission rate, constellation size, coding rate or a combination of these parameters [6].

Mobile radio links are exposed to multi-path fading due to the combination of randomly delayed reflected, scattered and diffracted components. In [7], a novel closed-form expression for achieving average channel capacity of a generalized selection combining rake receiver in Rayleigh fading is derived. A performance comparison of the capacity achieved with maximum ratio combination and rake receivers is also presented. Perera, Pollock and Abhayapala [8] has investigated the limits of information transfer over a fast *Rayleigh* fading Multi Input and Multi Output (MIMO) channel, where neither the transmitter nor the receiver is aware of the channel state information (CSI), except the fading statistics. Their work develops a scalar channel model in the absence of phase information for non-coherent *Rayleigh* fading and derives a capacity supremum with the number of receive antennas at any SNR using La-grange optimization. In [9], a unified L-branch equal gain combining (EGC) over generalized fading channels, such as Nakagmi-m, Rician, Hoyt or Weibull is presented. For each of these models, an exact closed-form expression is derived for the moments of the EGC output and SNR. Ekpenyong, Isabona and Umoren [10] extends [9] by deriving closed-form expressions for spectral efficiency of *Raleigh* fading channels, with EGC diversity for various adaptation policies. A methodology for computing the optimal cutoff SNR required for successful data transmission is also presented.

This paper is an extended version of [10]. The research improves on the bit error rate (BER) in the presence of SNR at various diversity levels using EGC and explores useful optimality conditions for achieving a robust system that minimizes degradation in the presence of fading.

3. BASIC DIVERSITY COMBINING METHODS

The collection of independently fading signal branches can be combined in a variety of ways to achieve a satisfactory (received) SNR. Since the chance of having two deep fades from two uncorrelated signals at any instant is rare, combining them can reduce the effect of the fades. The three most prevalent space diversity-combining techniques are selection diversity (SC), equal gain combination (EGC), and maximum radio combining (MRC). MRC co-phases the signal branches, weighs them according to their respective SNRs, and then computes their sum. MRC is the most complex combining technique, but this technique yields the highest SNR. A study of all of these diversity techniques is presented here.

3.1. Selection diversity

Selection diversity is the simplest of all the diversity schemes. It is based on the probability that the received signals rises above a threshold. An ideal selection combiner chooses the signal with the highest immediate SNR of all the branches, so the output SNR is identical to that of the best incoming signal and makes it available to the receiver at all times. However, multiple branches will increase the probability of having a better SNR at the receiver [11]. The algorithm for selective diversity combining is based on the principle of selecting the best signal among all of the signals received from different branches, at the receiver end. In the SC method for two antennas, as shown in Figure 1, the branch with the maximum voltage SNR is selected as the contributing received signal. For instance, branch one (1) *SNR* is chosen if it is larger than branch two (2) *SNR*. The weaker branch is not used.

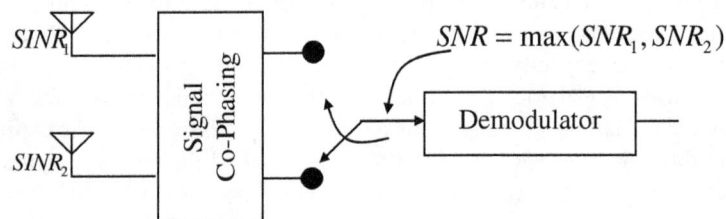

Figure 1. Two branch selection combining (SC) with equal noise power

In this method, the mean SNR of the selected signal is [12]:

$$< SNR >= \rho \sum_{m=1}^{M} \frac{1}{m} \cong \rho(C + \ln m + \frac{1}{2M})$$ (1)

where ρ is the mean branch SNR. The mean SNR improves logarithmically with the N number of branches.

3.2. Maximum Ratio Combining (MRC)

Maximum ratio combining gathers the information from all received branches for a multiple antenna system in order to increase the SNR. It employs different gains to each antenna to improve the signal to noise ratio for the combined signals. Maximum ratio combining can provide the diversity gain and array gain but it does not help in spatial multiplexing scenario [11]. The MRC method is shown in Figure 2, where each branch signal is first weighted by its received instantaneous voltage SNR, r_1M and r_2M, respectively. Thus the branch with the higher voltage SNR is weighted more than the branch with lower voltage SNR. The weighted signals are then co-phased and coherently summed. This method yields the highest instantaneous SNR possible using any linear combining technique. However, the implementation of the MRC is computationally expensive, as the weights need both amplitude and phase tracking of the channel response.

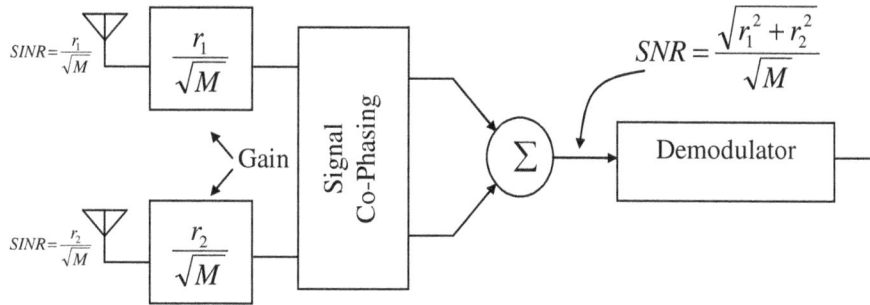

Figure 2. Two branch maximal-ratio combining (MRC) with equal noise power, source [13]

The mean SNR at the output of the combiner is [12]:

$$< SNR >= \sum_{m=1}^{M} \rho = M\rho \tag{2}$$

where ρ is the mean branch SNR. The mean SNR improves with the number of branches, M.

3.3. Equal Gain Combining (EGC)

EGC diversity receiver is of practical interest because of its reduced complexity relative to optimum maximal ratio combining scheme, while achieving near-optimal performance [11]. It is the accumulation of all the signals received in order to increase the available SNR at the receiver. The gain of all of the branches is set to a particular value that does not change which is in contrast to MRC. In the EGC method, the weights possess same constant magnitude, G, unlike the MRC, where the weights are based on the instantaneous voltages. As shown in Figure 3, the weighted signals are then co-phased and coherently summed. The noise power for the output voltage SNR is doubled.

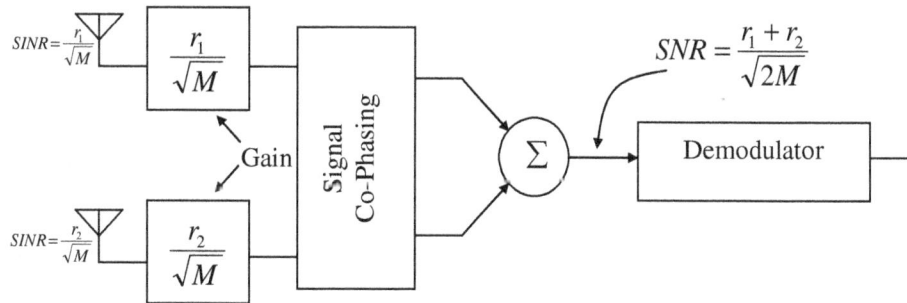

Figure 3. Two branch equal gain combining (EGC) with equal noise power, source [13].

The closed-form expression for the mean SNR at the output of the combiner is [12]:

$$< SNR >= \left[1 + (M-1)\frac{\pi}{4} \right] \rho \tag{3}$$

Here, the various diversity techniques which include Selection Combining (SC), Equal Gain Combining (EGC), and Maximum Ratio Combining (MRC), were simulated in MATLAB to compare the performance of the three techniques in terms of the complexity and improvement in SNR Rayleigh fading channel. From Figure 4, the plots of improved mean SNR as a function of number of diversity levels, M, show that MRC and EGC yield better gain than the SC method. For larger M, the MRC is about 1 dB higher than the EGC method. Hence, an improvement in the case of equal gain combining is comparable to that of maximal ratio combining.

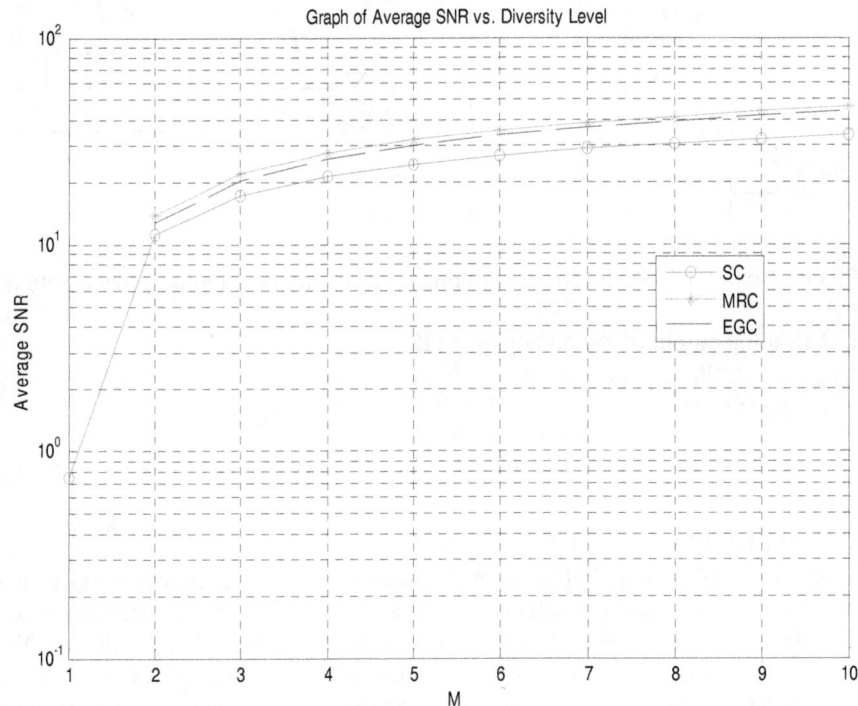

Figure 4. Graph of SNR vs. number of branches

In terms of the required processing, the selection combiner is the easiest – it requires only a measurement of SNR at each element and not the phase or the amplitude, i.e., the combiner need not be coherent. Both the maximal ratio and equal gain combiners, on the other hand, require phase information. The maximal ratio combiner requires accurate measurement of the gain too. But EGC is often used in practice because of its reduced complexity relative to the optimum MRC scheme. This is because the latter requires information on the fading amplitude in each signal branch while the former requires no such knowledge [14]. Thus, we focus more on the EGC performance in Rayleigh fading channel. We begin the investigation by exploring how the probability of error (BER) can be improved in the presence of interference with $M = 4$, 6, 8, 10 branches, using EGC. Three BER schemes: the Binary Phase shifts Keying (BPSK), Differential Phase shift Keying (DPSK) and Binary Frequency Phase shift Keying (BFSK) were used to specify different services in multimedia environment, i.e. data, voice and video channel. As can be seen from the plots in Figure 5-7, the three BERs improve as the diversity level increases.

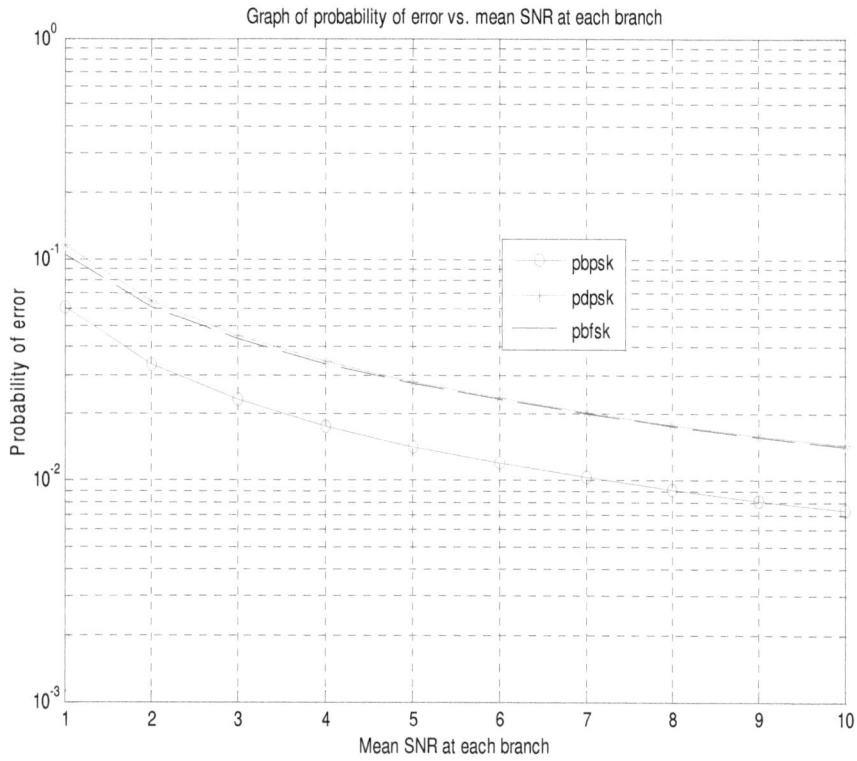

Figure 5. Graph of probability of error vs. SNR at each branch, for M=4

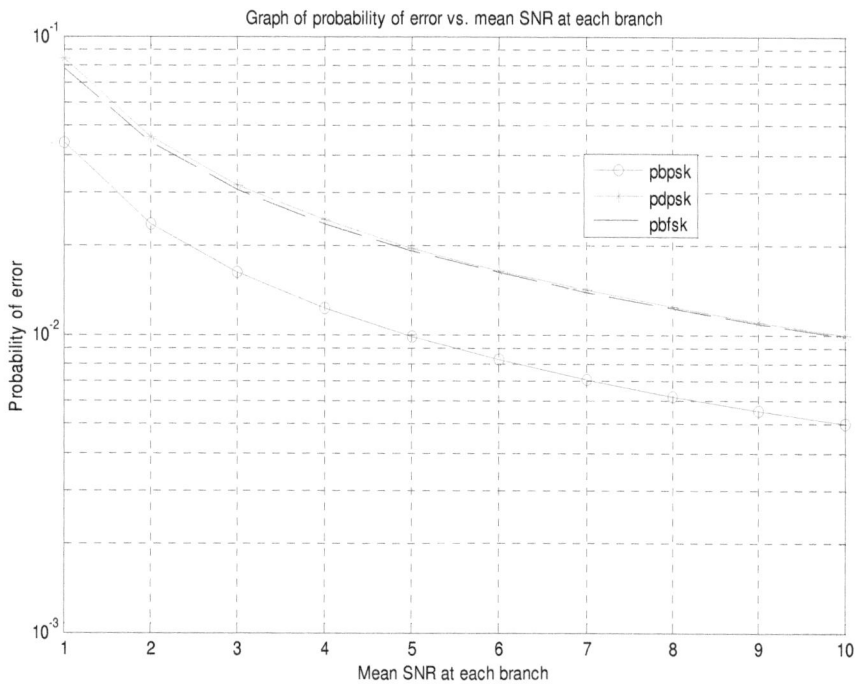

Figure 6. Graph of probability of error vs. SNR at each branch, for M=6

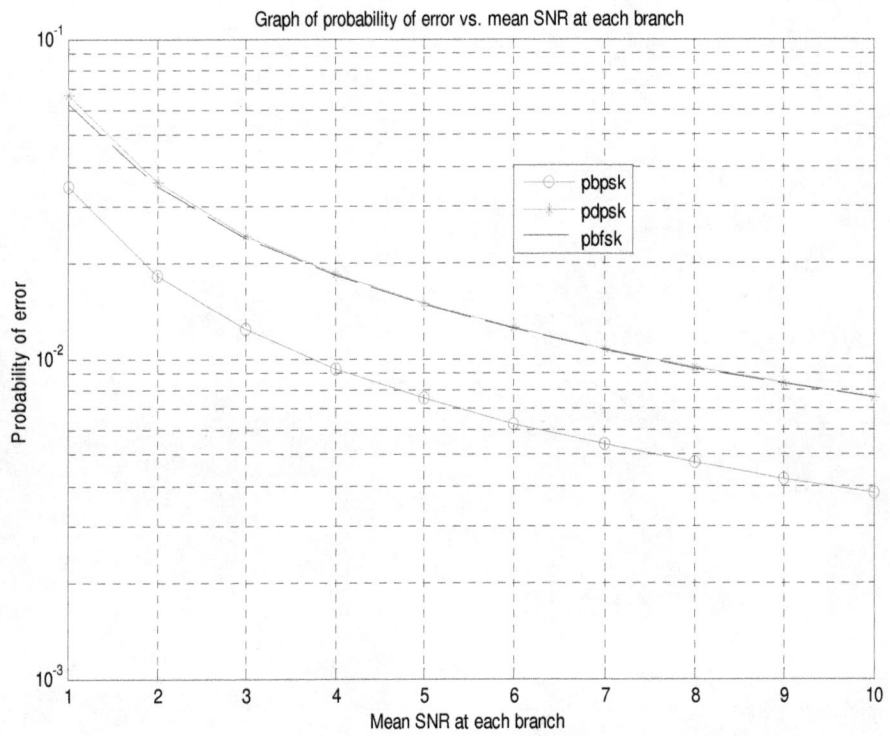

Figure 7. Graph of probability of error vs. SNR at each branch, for M=8

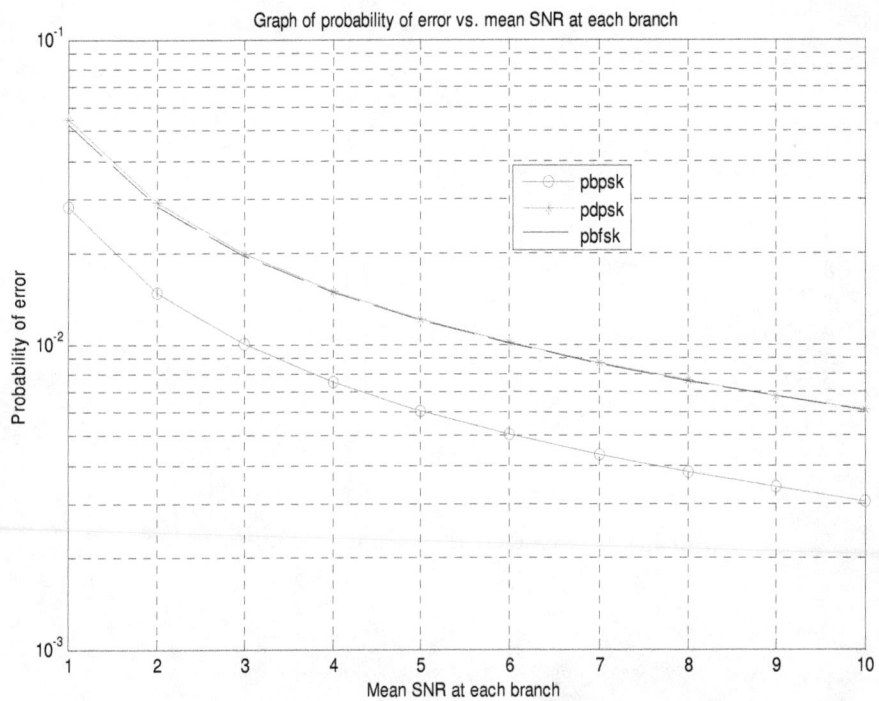

Figure 8. Graph of probability of error vs. SNR at each branch, for M=10

Next, we revisit the Shannon capacity efficiency of *Raleigh* fading channels, with EGC diversity under various adaptation policies. Though we have discussed these in [10], we repeat then here to enable readers follow the derivations of the system model and the purpose of our extension. The derivations as presented as follows:

3.3.1. Equal Gain Diversity Reception for Rayleigh Fading Channel

Given an average transmit power constraint, the channel capacity of fading channel with received SNR distribution, $P_\gamma(\gamma)$, and power and rate adaptation, C_p bits/s is given as [6]:

$$C_p = B \int_{\gamma_0}^{\infty} \log_2\left(\frac{\gamma}{\gamma_0}\right) \rho_\gamma(\gamma) \partial \gamma \tag{4}$$

where B(Hz) is the channel bandwidth and γ_0 is the cutoff level SNR below which data transmission is suspended. This cutoff must satisfy the following condition:

$$\int_{\gamma_0}^{\infty} \left(\frac{1}{\gamma_0} - \frac{1}{\gamma}\right) \rho_\gamma(\gamma) \partial \gamma = 1 \tag{5}$$

where $\rho_\gamma(\gamma)$ represents the *pdf* of the received signal amplitude for a *Rayleigh* fading channel with a *m*-branch EGC diversity and is given by

$$\rho_\gamma^{EGC}(y) = \frac{y^{m-1} e^{-y/\rho_x}}{\rho_x^m (m-1)!} \tag{6}$$

To achieve the capacity in equation (4), the channel fading level must be tracked both at the receiver and transmitter and the transmitter has to adapt its power and rates for excellent channel conditions (i.e., γ is large), by maintaining lower power levels and rates for unfavourable channel conditions (γ is small). Since no data is sent when $\gamma < \gamma_0$, the optimal policy suffers a probability of outage, P$_{out}$, equivalent to the probability of no transmission, given by:

$$P_{out} = \int_0^{\gamma} \rho_\gamma(\gamma) d\gamma = 1 - \int_{\gamma_0}^{\infty} \rho_\gamma(\gamma) d\gamma \tag{7}$$

Substituting equation (6) into (5), and simplifying same, we observe that γ_0 must satisfy

$$\Lambda^{(C)}\left(m, \frac{\gamma}{\rho_x}\right) \frac{-\gamma_0}{\rho_x} \bullet \Lambda^{(C)}\left(m-1, \frac{\gamma_0}{\rho_x}\right)$$
$$= (m-1)! \frac{\gamma_0}{\rho_x^{m-1}} \tag{8}$$

where $\Lambda^{(C)}(\alpha, x) = \int_x^{\infty} t^{\alpha-1} e^{-t} dt$ is the complementary incomplete gamma function [14]. Let $v = \gamma/\rho_x$, and $f(v)$ is defined as:

$$f(v) = \Lambda^{(C)}(m, v) - v\Lambda^{(C)}(m-1, v) - \frac{v}{\rho_x^{m-2}}(m-1)! \tag{9}$$

Substituting equation (6) into (4), we have

$$\frac{C_p^{EGC}}{B} = \frac{1}{\rho_x^m (m-1)} \int_{\gamma_0}^{\infty} \log_2 \left(\frac{\gamma}{\gamma_0} \right) \gamma^{m-1} e^{-\gamma/\rho_x} \, d\gamma \tag{10}$$

3.3.2. Flat Fading and Frequency Selective Fading

As the carrier frequency of a signal is varied, the magnitude of change in amplitude also varies. The coherence bandwidth measures the minimum separation in frequency after which both signals experience uncorrelated fading [16]. In flat fading, the coherence bandwidth of the channel is usually higher than the bandwidth of the signal. Therefore, all frequency components of the signal will experience the same magnitude of fading.

In frequency selective fading, the coherence bandwidth of the channel is lower than the bandwidth of the signal. Different frequency components of the signal therefore experience de-correlated fading, since different frequency components of the signal are independently affected.

3.3.3. Optimal Rate Adaptation to Channel Fading with Constant Transmit Power Policy

With optimal rate adaptation to channel fading at a constant transmit power, the channel capacity, C_0 (bits/s) becomes [4]:

$$C_o = \beta \int_0^{\infty} \log_2 (1 + \gamma) P_\gamma (\gamma) d\gamma \tag{11}$$

Substituting equation (6) into (11), we obtain

$$\frac{C_o^{EGC}}{B} = \log_2 (e) \frac{1}{\rho_x^m (m-1)!} \times \int_0^{\infty} \log_e (1 + \gamma) e^{-\xi\gamma} \gamma^{m-1} d\gamma$$

$$= \log_2 (e) \frac{1}{\rho_x^m (m-1)!} I_m (\xi) \tag{12}$$

where

$$I_m (\xi) = \int_0^{\infty} t^{n-1} \log_e (1+t) e^{-\xi t} dt, \ \xi > 0$$

m represents the diversity levels
ρ_m^x is the average SNR

Using the result of $I_m (\xi)$, we can rewrite equation (11) as:

$$\frac{C_o^{EGC}}{B} = \log_2 (e) e^{\xi} \Lambda^C (-k, \xi) \tag{13}$$

and equally express in the form of a Poisson distribution as [17]:

$$\frac{C_o^{EGC}}{B} = \log_2(e)(P_M(-\xi)E_1(\xi)) + \sum_{k=1}^{M-1} \frac{P_k(\xi)P_{M-k}(-\xi)}{k} \qquad (14)$$

where $P_k(\xi)$ is given by $P_k(\xi) = e^{-\xi}\sum_{j=0}^{k-1}\frac{\xi_j}{j!}$ and $E_1(\xi) = \int_1^\infty \frac{e^{-\xi t}}{t}dt$.

3.3.4. Channel Capacity with Fixed Rate Policy

The channel capacity when the transmitter adapts its power to maintain a constant SNR at the receiver or inverts the channel fading is also investigated in [4]. This technique uses fixed rate modulation and a fixed code sign, since the channel after channel inversion appears as a time invariant AWGN channel. As a result, channel inversion with fixed rate is the least complex technique to implement, assuming good channel estimates are available at the transmitter and receiver. With this technique, the channel capacity of an AWGN channel is given as [4]:

$$C_c = B\log_2\left(1 + \frac{1}{\int_0^\infty (\rho_\gamma(\gamma)/\gamma)\,d\gamma}\right) \qquad (15)$$

Inverting channels with fixed rate suffers large capacity penalty relative to other techniques, since a large amount of the transmitted power is required to compensate for the deep channel fades.

The capacity with truncated channel in varied and fixed rate policies $C_t(bits/s)$ becomes:

$$C_t = B\log_2\left(1 + \frac{1}{\int_{\gamma_0}^\infty (\rho_\gamma(\gamma)/\gamma)\,d\gamma(1-P_{out})}\right) \qquad (16)$$

where P_{out} is given in equation (7). We then obtain the capacity per unit bandwidth with EGC diversity for channel inversion with fixed rate policy (total channel inversion) as C_c/B, by substituting equation (6) into equation (16) giving:

$$\frac{C_c}{B} = \log_2\left[1 + \frac{\rho_x^m\Gamma(m)}{\int_0^\infty \gamma^{m-2}e^{\gamma/\rho_x}\,d\gamma}\right] \qquad (17)$$

Now, substituting $t = \gamma/\rho_x^m$ and $dt = d\gamma/\rho_x^m$ into equation (15), yields,

$$\frac{C_c^{EGC}}{B} = \log_2\left[1 + \frac{\rho_x\Gamma(m)}{\int_0^\infty t^{m-2}e^{-t}dt}\right] = \log_2[1 + (m-1)\rho_x] \qquad (18)$$

The capacity of this policy for a *Rayleigh* fading channel is the same as the capacity of an AWGN channel with equivalent $SNR = (m-1)\rho_x$. Truncated channel inversion improves the capacity in equation (16) at the expense of the outage probability, P_{out}^{EGC}. The capacity per unit bandwidth of truncated channel inversion with EGC diversity C_t^{EGC}/B, is obtained by substituting equation (6) into equation (14). Thus,

$$\frac{C_t^{EGC}}{B} = \log_2\left[\frac{1+\rho_x^m(m-1)!}{\int_{\gamma_0}^{\infty}\gamma^{m-2}e^{-\gamma/\rho_x}}\partial\gamma\right]\times\left(1-P_{out}^{EGC}\right) \qquad (19)$$

where P_{out}^{EGC} is given in equation (6). Substituting $t = \gamma/\rho_x$ and $dt = d\gamma/\rho_x$ into equation (17), gives:

$$\frac{C_t^{EGC}}{B} = \log_2\left[1+\frac{\rho_x(m-1)}{\int_{\gamma_0/\rho_x}^{\infty}t^{m-2}e^{-t}dt}\right]\bullet\int_{\gamma_0}^{\infty}P_{\gamma}^{EGC}(\gamma)\partial\gamma \qquad (20)$$

Substituting $t = \gamma/\rho_x$ and $dt = d\gamma/\rho_x$ into equation (18), we arrive at:

$$\frac{C_t^{EGC}}{B} = \frac{1}{\Gamma(m)}\log_2\left[1+\frac{\rho_x\Gamma(m)}{\Lambda^{(c)}(m-Q,\mu)}\right]\times\int_{\gamma_0/\rho_x}^{\infty}t^{m-1}e^{-t}dt$$

$$= \frac{\Lambda^{(c)}(m,\gamma_0/\rho_x)}{(m-1)!}\log_2\left[1+\frac{\rho_x(m-1)!}{\Lambda^{(c)}(m-Q,\mu)}\right] \qquad (21)$$

4. DISCUSSION OF RESULTS

Using the three channel capacity schemes we established relationships between the model parameters through extensive computer simulation. We continuously fine-tuned the parameters/characteristics behaviour of the system until an optimum solution was achieved. Table 1 shows the sample input used during the simulation:

Table 1. Sample input parameters and values

Parameter	Value
SNR (dB) (P_x)	5, 10, 15
Diversity level (M)	1-10

Figures 9-11 show the effect of diversity level on channel capacity at different average SNR values (SNR= 5, 10, 15), for the various transmission policies. We observe that channel capacity increases with the diversity level and improves for all the policies. This upper bound on the amount of information that can be reliably transmitted over communication channel is well maximized based on the input distribution, for channel inversion with fixed rate policy and truncated channel inversion, respectively, as can be seen from the plots. Also, the bits are satisfactorily controlled and do not exceed the Shannon bound, thus reducing the number of errors. This confirms the notion that if we attempt to send bits faster than this rate, the error rate will rise beyond a negligible value, thus causing large negative effects on the communication system.

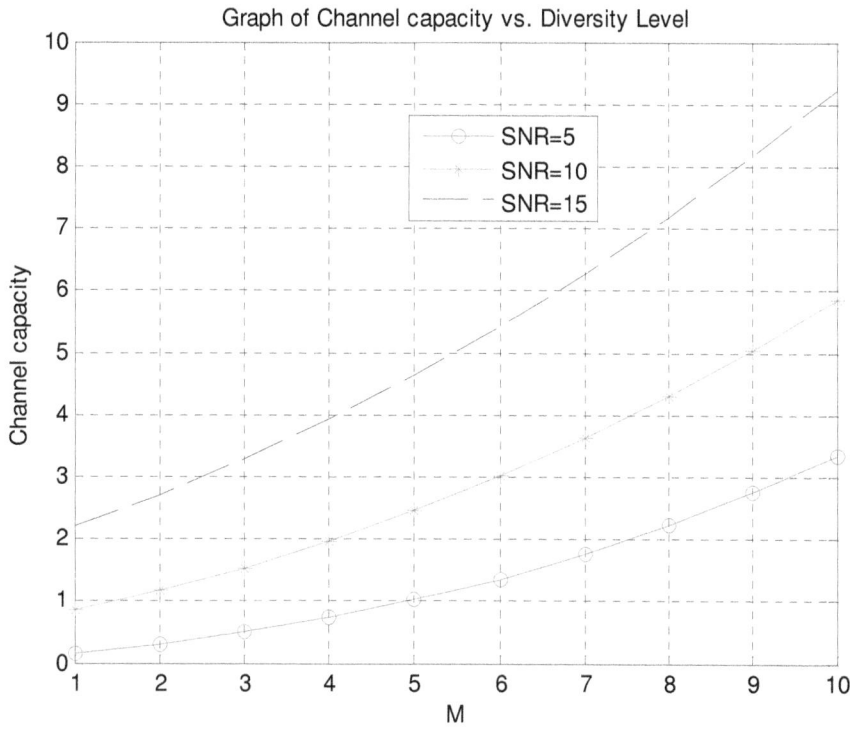

Figure 9. Graph of channel capacity vs. diversity level, for Channel capacity with optimal rate adaptation policy, at constant transmit power

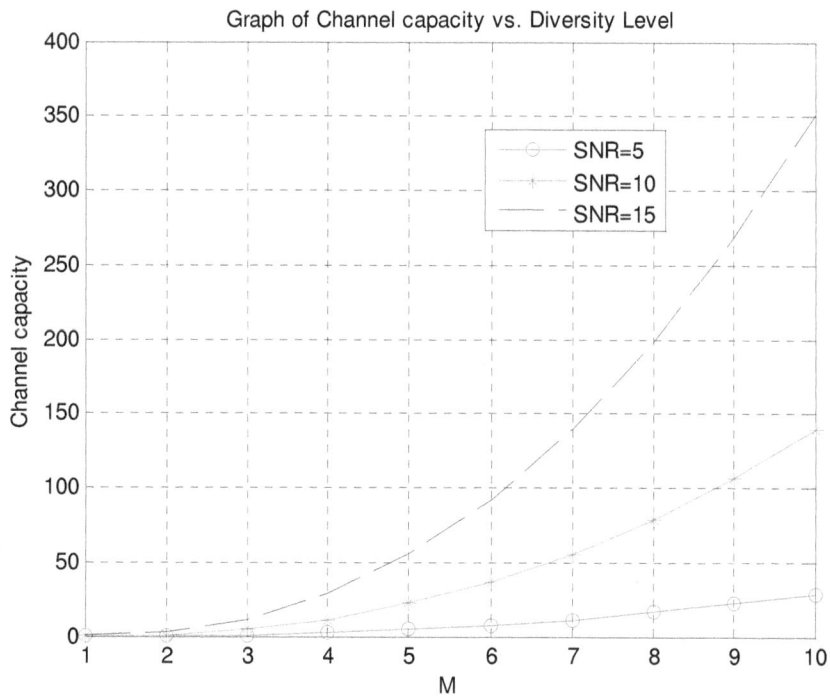

Figure 10. Graph of channel capacity vs. diversity level, for channel capacity with EGC diversity and channel inversion with fixed rate policy

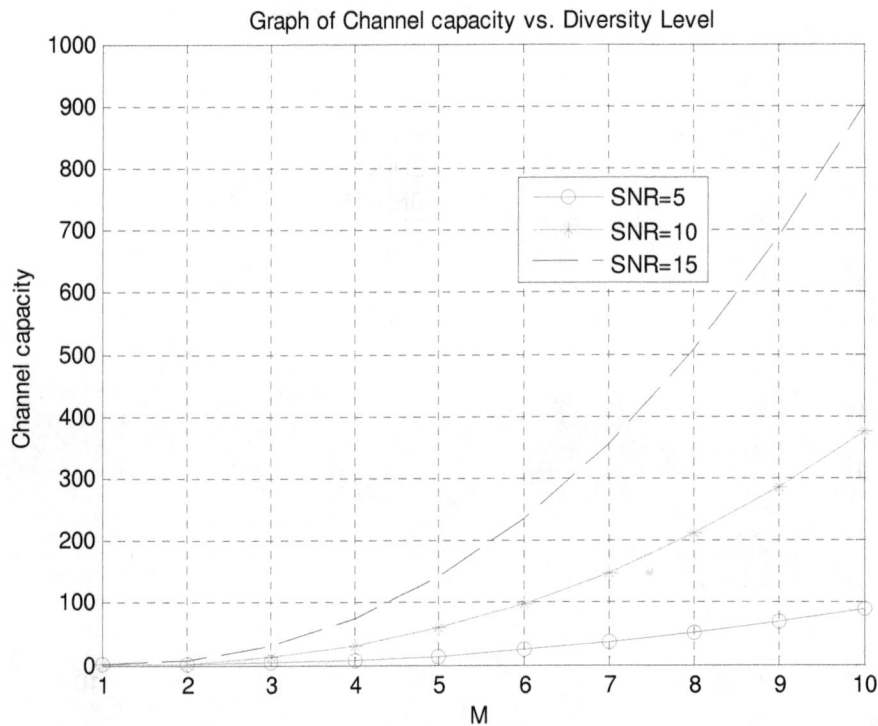

Figure 11. Graph of channel capacity vs. diversity level, for channel capacity with EGC diversity and truncated channel inversion

5. CONCLUSION

We have studied the different diversity combining techniques, with special interest on the estimation of the capacity of equal gain combining (EGC) under a multi-path fading channel. In particular, we obtained closed-form expressions for the channel capacities of the various adaptation policies used in conjunction with diversity combining. In order to evaluate the impact of the three adaptation policies on the diversity schemes, we simulated the schemes under ideal communication conditions using a robust simulation software. From the simulation results, channel inversion yields the best system performance compared to other diversity schemes. Hence, to maintain the desired level of productivity and contend with the ever increasing users' capacity, mobile network operators must conserve, share and manage available bandwidth efficiently.

Optimal power and rate adaptation yields a small increase in capacity over just optimal rate adaptation, and this small increase in capacity diminishes as the average received SNR and/or the number of diversity branches increases. In addition, channel inversion suffers the largest capacity penalty relative to the two other policies. However, this capacity penalty diminishes with increasing diversity.

REFERENCES

[1]   Rappaport, T. S. (2002). *Wireless Communications – Principles and Practice*, Second Edition, Prentice Hall PTR.

[2]   Brennan, D. (2003). Linear Diversity Combining Techniques. In Proceedings of IEEE. 91(2):331-356.

[3]   Alouini, M. and Goldsmith, A. (1999). "Capacity of Rayleigh Fading Channels under Different Adaptive Transmission and Diversity Combining Techniques". *IEEE Transactions on Vehicular Technology*. 48(4): 1165-1181.

[4]   Goldsmith, A. and Varaiya, P. (1997). "Capacity of fading channels with channel side Information". *IEEE Transmissions on Information Theory*: 43(6): 1986-1992.

[5]   Alouini, M. and Goldsmith A. (1997), Capacity of Nakagami Multipath Fading Channels. In proceedings of the IEEE Vehicular technology conference VTC'97 Phoemix, A2: 358-3620.

[6]   Chua, S. and Goldsmith, A. (1996). Variables rate variable Power M-QAM for fading channels. In Proceedings of the IEEE Vehicular Technology Conference: 815-819.

[7]   Sagias, N., Varzakas, P., Tombras, G., and Karagiannidias, G. (2004). "Average Channel Capacity for Generalized-Selection Combining RAKE Receivers". *Euro Transactions on Telecommunication*. 15: 497-500.

[8]   Perera, R., Pollock, T. and Abhayapala, T. (2005). Non-coherent Rayleigh Fading MIMO channels: Capacity Supremum. In Proceedings of Asia Pacific Conference on Communication (APCC), Perth, Australia: 72-76.

[9]   Karagiannidis, G., Sagias, N. and Zogas, O. (2005). Analysis of M-QAM with Equal-gain Diversity over Generalized Fading Channels. In IEEE Proceedings on Communication. 152(1): 69-74.

[10]   Ekpenyong, M., Isabona, J. and Umoren, I. (2012). On the Estimation Capacity of Equal Gain Diversity Scheme Under Multi-path Fading Channel. In Proceedings of Fourth International Conference on Network and Communications. ***

[11]   Howard Bonds III, System Performance in Fading Channel Environments, M.S.E.E, May 2003

[12]   Janaswamy, R. *Radiowave Propagation and Smart Antennas for Wireless Communications*, Kluwer Academic Publishers, Norwell, MA, 2001.

[13]   Dietze, K. (2001). Analysis of a Two-Branch Maximal Ratio and Selection Diversity System with Unequal Branch Powers and Correlated Inputs for a Rayleigh Fading Channel. Master's Thesis, Virginia Polytechnic Institute and State University.

[14]   Mohammed, A. K and Widad, I. (2010). Performance Diversity Techniques for Wireless Communication System, Journal of Telecommunication, 4(2): 1-6.

[15]   Gradshteyn, I. and Ryzlink, I. (1994). *Table of integrals, series, and products*. 5th Edition. San Diego, CA: Academic Press.

[16]   Tse, D. and Viswanath, P. (2005). *Fundamentals of Wireless Communication*, Cambridge University Press.

[17]   Gunther C. (1996). "Comment on Estimate of Channel Capacity in Rayleigh Fading Environment". IEEE Transactions of Vehicular Technology, 45:401–403.

17

DESIGN AND IMPLEMENTATION OF LOW LATENCY WEIGHTED ROUND ROBIN (LL-WRR) SCHEDULING FOR HIGH SPEED NETWORKS

Zuber Patel[1] and Upena Dalal[2]

[1]Department of Electronics Engg., National Institute of Technology, Surat, India

[2]Department of Electronics Engg., National Institute of Technology, Surat, India

ABSTRACT

Today's wireless broadband networks are required to provide QoS guarantee as well as fairness to different kinds of traffic. Recent wireless standards (such as LTE and WiMAX) have special provisions at MAC layer for differentiating and scheduling data traffic for achieving QoS. The main focus of this paper is concerned with high speed packet queuing/scheduling at central node such as base station (BS) or router to handle network traffic. This paper proposes novel packet queuing scheme termed as Low Latency Weighted Round Robin (LL-WRR) which is simple and effective amendment to weighted round robin (WRR) for achieving low latency and improved fairness. Proposed LL-WRR queue scheduling scheme is implemented in NS-2 considering IEEE 802.16 network [1] with real time video and Constant Bit Rate (CBR) audio traffic connections. Simulation results show improvement obtained in latency and fairness using LL-WRR. The proposed scheme introduces extra complexity of computing coefficient but its overall impact is very small.

KEYWORDS

Scheduling, WRR, LL-WRR, fairness, latency

1. INTRODUCTION

The phenomenal growth in real time services such as interactive voice & video poses challenge in meeting end to end QoS [2] requirement. Unlike non-real time data services, these real time applications have stringent performance requirements. Keeping this in mind, many schemes have been proposed by researchers for efficient packet queuing and scheduling. The primary job of queuing and scheduling is to treat different traffic classes with variable degree of priority to provide performance guarantee for range of different traffic types and profiles. They determine the order in which packets from different service classes are processed and served, hence it dictates resource allocation to different connections.

The scenario considered here consists of packet queues of various connections (or sessions) waiting for transmission through a single output port of network node. Scheduler component of network node schedules packets based on some policy so as to achieve requirements of each connection such as minimum reserved transmission rate (MRTR), latency, jitter and fairness. It is desirable to have low complexity in the implementation of scheduler to provide QoS in high speed converged networks. A queue scheduling scheme may not possess all of the above desirable scheduling properties instead offers subset of them. For example, weighted fair queuing (WFQ) [3] and its variant worst-case fair weighted fair queuing (WF2Q) [4] are having good delay and fairness properties but have high implementation complexity. Self clocked fair queuing (SCFQ) [5] uses the virtual finish time of the packet that is currently being transmitted as the system virtual time. As a result, the complexity of computing the system virtual time [6] of SCFQ is O(1) but delay increases linearly with increase in number of sessions. RPS based schemes (such as FFQ [7], SPFQ and MD-SCFQ) offer low complexity at the expense of degradation in fairness. Many schemes such as WFQ and SPFQ fail to provide stable latencies for real time traffic. This issue is addressed in [8] to guarantee low and stable latency for real time flows.

This paper proposes modification to WRR named as LL-WRR that improves *latency* and *fairness* of real time services with very small impact on complexity and data rate. LL-WRR is a simple and lightweight method of controlling *Round Robin length* by changing integer weights of connections while keeping the simplicity of WRR. The Round Robin length signifies the sum of packets to be scheduled from all queues in single round. To obtain integer weight, fractional weight of each connection is multiplied by constant integer in classical WRR. Thus, the sum of these integer weights i.e. Round Robin length is fixed and does not scale well with variation in number of connections. In LL-WRR scheme, instead of multiplying constant integer, a *co-efficient* γ is multiplied to fractional weights. The coefficient γ is function of number of connections present in network and it is made to decrease as number of connections increases. This eventually reduces Round Robin length and latency remains low. The co-efficient γ (hence Round Robin length) is computed only at the beginning of each WRR cycle rather than every packet arrival or departure. This keeps complexity of LL-WRR low.

The rest of the paper is organized as follows. Brief idea of conventional WRR, list based interleaved WRR and Multiclass WRR is reviewed in Section 2. Section 3 presents proposed LL-WRR scheme with the analysis on latency, rate and fairness properties. Section 4 discusses simulation results obtained in NS-2. Finally, the conclusion remarks are given in Section 5.

2. RELATED WORK

Generally speaking, packet scheduling algorithms can be divided into two categories (1) Timestamp based (2) Round Robin based. Timestamp-based algorithms have provably good delay and fairness properties, but generally need to sort packet deadlines, and therefore suffer from complexity logarithmic in the number of flows N. Generalized Processor Sharing (GPS) [3] (also called Fluid Fair Queuing) is considered the ideal time stamped scheduling discipline that achieves perfect fairness and isolation among competing flows. However, the fluid model assumed by GPS is not amenable to a practical implementation. However, GPS acts as a reference for other scheduling disciplines in terms of delay and fairness. Practical timestamp schedulers try to emulate the operation of GPS by computing a timestamp for each packet. Packets are transmitted in according to their timestamps. WFQ and WF2Q are examples of time stamped schedulers. WFQ is packet-by-packet equivalent of GPS. WFQ exhibits some short-term unfairness which is addressed by the WF2Q.

Round-robin-based algorithms achieve O(1) complexity by eliminating time stamping and sorting. The simplicity of these algorithms can be useful for traffic scheduling in very high speed networks. They support fair allocation of bandwidth, but unable to provide good delay bounds. Most basic round robin type scheduler for differentiated services network is WRR. WRR assures fraction of output link bandwidth to each service queue by assigning appropriate weight. The deficit round robin (DRR) is modification of WRR which takes into account packet size for scheduling. In following sections, we shall discuss and analyze few WRR based methods namely conventional WRR, List based interleaved WRR and Multiclass WRR.

2.1. Conventional WRR

The WRR is simple Round Robin based scheduling algorithm used in packet-switched networks with static weight assigned to connections' queues. It cycles through queues transmitting amount of packets from each queue as per its weight (Figure 1) thus guaranteeing each connection a fraction of output link bandwidth. It also ensures that lower priority queues never starved for long time for buffer space and output link bandwidth. It has processing complexity of O(1) which make it feasible to high speed interfaces in both core and at the edge of network. The primary limitation of WRR queuing is that it provides correct percentage of bandwidth to each service class only if all of the packets in all queues are of same size or when mean packet size is known in advance.

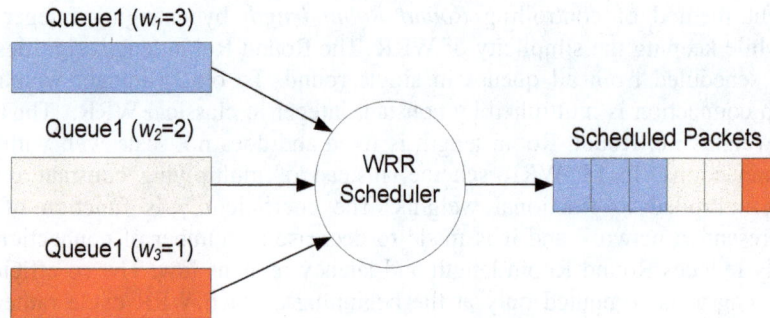

Queue1 (w_1=3)

Queue1 (w_2=2)

WRR Scheduler

Scheduled Packets

Queue1 (w_3=1)

Figure 1. Operation of Conventional WRR

WRR scheduling is based on assigning fraction weight ϕ_i to each service queue such that sum of weight of all service queues is equal to one.

$$\sum_{i=1}^{N} \phi_i = 1 \qquad (1)$$

Since weight is *fraction* and we want to determine number of integer packets to be served from each queue, the fraction weight is multiplied by proper constant integer M. The product is rounded off to nearest larger integer to obtain *integer weight* w_i. This integer weight value of each queue specifies number of packets to be serviced from that queue. The total sum of these counter values is referred to as round robin length. The integer weight of ith queue is

$$w_i = \lceil \phi_i * M \rceil \qquad (2)$$

The sum of existing N active connections is defined as round robin length W and is given by

$$W = \sum_{i=1}^{N} w_i = M \qquad (3)$$

Assume that the rate of outgoing link is r, and the rate offered to i^{th} connection is

$$ri = \frac{w_i}{W} r \qquad (4)$$

Let us understand the effect of increasing number of connections on the rate. As number of connection N increases, the equality of Equation (1) tells that individual weight of connection ϕ_i decreases and this reduces w_i. Since sum of all weight remains constant, W remains unchanged and hence as per Equation (4) rate of that connection decreases.

The latency θ_i of any connection i defined in [9] and is adopted in our analysis. For particular scheduling algorithm, parameters such as transmission rate of output link, allocated rates and number of connections may influence latency. We determine worse-case latency for connection i for conventional WRR scheduler. Assume that there are N connection queues being backlogged and scheduler is currently serving w_ith packet from ith connection. Since cycle length is W, there could be as many as W-w_i packets to be served from other N-1 queues before (w_i+1)th packet from connection queue i is served. Therefore worst-case latency for ith connection is

$$\theta_{i,WRR} = (W - w_i)\frac{L_i}{r} = W(1 - \phi_i)\frac{L_i}{r} \tag{5}$$

where L_i is the maximum length of packet that belongs to ith connection. This worst-case latency increases as ϕ_i decreases with increase in number of connections. Hence it has inefficient latency tuning characteristics. To compute *total latency* experienced by a packet, queuing latency should be added to Equation (5).

The proportional fairness $\eta_{PF} = 1$ since a connections i (j) can lead the other connection j (i) at the most by w_i (w_j) packets. To measure *worst-case fairness*, a metric called Worst case Fair Index (WFI) is defined in [4] to characterize fair queuing servers. A server is said to guarantee a WFI of C_i for connection i, if for any time t the delay of a packet arriving at t is bounded

$$d_i < a_i + \frac{Q_i(t)}{r_i} + C_i \tag{6}$$

where $Q_i(t)$ is the queue size of connection i at the packet arrival time t and C_i is called worst case fair index for connection i. Suppose a new packet of connection i arrives at time t when the server has just crossed i, and suppose the backlog of connection i at time t (denoted by $Q_i(t)$) is multiple of w_i. Then, this packet departs after a maximum of time of $[Q_i/r_i + (W + w_i + 1)L_i/r]$. Thus WFI of WRR scheduler is given by

$$C_{i,WRR} = d_i - a_i - \frac{Q_i(t)}{r_i}$$
$$= (W - w_i + 1)\frac{L_i}{r} \tag{7}$$

As the number of connections increases, w_i decreases and hence WFI increases. So increase in number of connections on scheduler degrades WFI.

2.2. List based interleaved WRR

In list based WRR scheme [10], instead of serving w_i packets from ith connection in single visit, the service is distributed evenly over the entire Round Robin cycle. Scheduler visits queues of connections as per the "*service list*" maintained by it. The number of times connection i appears in the service list is proportional to its weight w_i, but these appearance are not necessarily consecutive as in conventional WRR. The service list is updated only at the time of new connection establishment or connection termination.

In order to form service list, we create M ($=max_{i=1 to N} (w_i)$) slots in service list and each slot contains entries of indices of connections. A connection i will have w_i entries in service list evenly distributed across all slots. The total length of service list is $W = \sum_{i=1}^{N} w_i$ is called Round Robin length. Scheduler parses this service list and determines queues to be serviced.

Consider any two connections i and j with $w_j \geq w_i$. In list based WRR scheduling, the connection j can lead i by at the most $w_j - w_i + 1$ packets in any partial Round Robin cycle. Let us assume that $S_i(t, t+\tau)$ and $S_j(t, t+\tau)$ are services offered to connections i and j respectively by server during interval $(t, t+\tau)$. Then as per definition given by [5] the proportional fairness η_{PF} is the difference in the normalized services offered to i and j. Because of cyclic nature of scheduling, the maximum normalized service by which j can lead i, is the same as maximum normalized service by which i can lead j. In other words,

$$\left| \frac{S_i[t, t+\tau]}{w_i} - \frac{S_j[t, t+\tau]}{w_j} \right| \leq \frac{w_j - w_i + 1}{w_j} \tag{8}$$

Hence, for this list based scheme the proportional fairness is

$$\eta_{PF} = \frac{w_j - w_i + 1}{w_j} \tag{9}$$

Since the values of w_i and w_j are normally larger than 1, η_{PF} of this scheme is smaller than conventional WRR and hence it has better proportional fairness. Its latency value $(W - w_i) = W(1 - \phi_i)$ suggests that this scheme also lacks efficient latency tuning.

2.3. Multiclass WRR

This scheme offers scheduling properties similar to WFQ based schemes. Multiclass WRR [10] has efficient tuning characteristics and it is worst-case fair. To get the initial grasp of Multiclass WRR, consider M classes from ϕ_1 to ϕ_M containing N_1 to N_M connections respectively and all connections have unity weight. Also, let W_1 to W_M represents maximum length of Round Robin cycle of class ϕ_1 to ϕ_M respectively in increasing order of size. Multiclass WRR works on *minicycle* which is set to W_1 visits. The scheduler operates by embedding smaller Round Robin cycles within a minicycle. In every minicycle, all connections of class ϕ_1 are always visited. After this, the connections in subsequent classes are visited from the leftover visits from previous classes. That is to say, connections in any class ϕ_m are visited from leftover visits from classes ϕ_1, ϕ_2, ... ϕ_{m-1} in a minicycle. If the fraction of the output link bandwidth assigned to connection i is ϕ_i then from equality $\sum_{i=1}^{M} \phi_i \leq 1$ following condition holds:

$$\sum_{m=1}^{M} \frac{N_m}{W_m} \leq 1 \tag{10}$$

The operation of Multiclass WRR is shown pictorially in Figure 2 Notice that the minicycle may or may not end at the boundary of class. Besides, it may require one or more visits of minicycle to serve all connections of any class after ϕ_1. In the scenario of Figure 2, the first minicycle terminates after visiting the $(N_1+\beta_1)^{th}$ connection. During the second round minicycle when the server crosses the class ϕ_1 boundary, it jumps to visit the $(N_1+\beta_1+1)^{th}$ connection. The second minicycle ends at the end of ϕ_2. The third minicycle ends when the last connection in class ϕ_2 is visited.

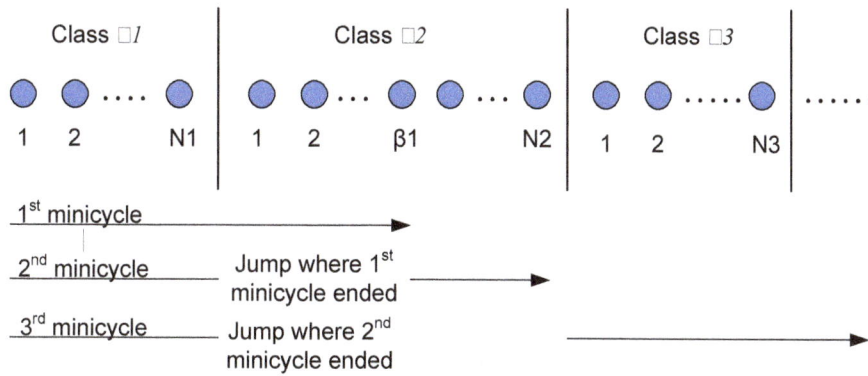

Figure 2. Operation of Multiclass WRR with minicycles

It can be shown that Multiclass WRR has near optimal proportional fairness over any interval during which both connections i and j are continuously back-logged. In addition, it has efficient latency tuning characteristics and it is worst-case fair.

Many other modifications to WRR have been suggested in research literature. Improvement in average delay and throughput with large service round is discussed in [11] whereas [12] suggests a *revenue based* criterion is to adjust weight to allocate resources in optimum way that maximizes total revenue. In other work [13], negative deficit weighted round robin (N-DWRR) is proposed which schedules packets based on credit, negative credit and packet size to improve the bandwidth utilization rate without increasing the total latency.

3. PROPOSED LOW LATENCY WRR (LL-WRR) SCHEME

Proposed LL-WRR scheme is simple extension to WRR which represents tradeoff between *rate* and *worst-case latency*. We first discuss the basic concept of proposed scheme. In real networks, packets of a connection experience queuing latency and scheduling (processing) latency. In section 2.1, we have seen that WRR scheduling latency increases with number of connections linearly. The queuing latency, however, increases nonlinearly with traffic load [14]. Assuming the same arrival rate of packets on each connection, the traffic load is directly proportional to the number of connections. Assuming M/M/1 queuing with Poisson's arrival at rate λ and service rate μ and traffic load on i^{th} queue is ρ_i, the average queuing latency T_i of ith connection is given by little's formula

$$T_i = \frac{\rho_i}{\mu(1 - \rho_i)} \tag{11}$$

The *total latency* varies as shown Figure 3(a) with number of connections N. It rises slowly initially and then increases rapidly as N becomes very large. In our proposal, to improve worst-case latency, a coefficient γ is multiplied to weight ϕ_i of each connection and is made to vary with N. Since latency is proportional to γ, we can keep latency low by decreasing γ with increase in N. To counteract rise in total latency (Figure 3(a)), we reduce co-efficient slowly initially for smaller value of N and then decrease rapidly for large value of N. In other words, co-efficient must vary as shown in Figure 3(b). The function that satisfy this characteristics is given as

$$\gamma = \gamma_0 \sqrt{1 - \eta (N/N_{max})^2} \qquad (12)$$

where γ_0 is appropriate integer constant that decides Round Robin length for a given number of connections N. The constant fraction η (having value less than 1) decides the value of γ when number of connections N reaches to maximum value i.e. N_{max}. The integer weight of ith connection w_i is obtained by ceiling the value of product $\gamma\Phi_i$. The corresponding cycle length W is sum of w_i. The scheduler latency of our scheme is obtained by substituting these values of w_i and W in Equation (5) and then total latency $\theta_{i,LL\text{-}WRR}$ can be obtained by adding queuing latency of Equation (11) to scheduling latency.

$$\theta_{i,LL-WRR} = \left[\left(\sum_{i=1}^{N} \lceil \gamma \phi_i \rceil \right) - \lceil \gamma \phi_i \rceil + \frac{\rho_i}{\mu(1-\rho_i)} \right] \qquad (13)$$

Similarly, rate of ith connection is obtained from Equation (4) by substituting new values of w_i and W

$$r_i = \frac{\lceil \gamma \phi_i \rceil}{W} r \qquad (14)$$

From Equation (13) we can surmise that as number of connections increases, the total latency tends to remain very low. This is because as number of connections increases, γ reduces rapidly and the term $\sum_{i=1}^{N} \lceil \gamma \phi_i \rceil$ reduces. Thus increase in $\Theta_{i,LL\text{-}WRR}$ due to reduction in $\lceil \gamma \phi_i \rceil$ and increase in traffic load ρ_i is compensated by decrease in $\sum_{i=1}^{N} \lceil \gamma \phi_i \rceil$. The Equation (14) suggests that data rate performance is negligibly affected at low value of N. But as N increases to large value, even if W decreases, significant reduction in $\lceil \gamma \phi_i \rceil$ reduces ratio and hence the data rate r_i. This demonstrates the trade-off between latency and data rate. This scheme also improves *worst-case fairness*. From Equation (7), we may write WFI for LL-WRR as

$$C_i = (W - w_i + 1)\frac{L}{r} = \left(\sum_{i=1}^{N} \lceil \gamma \phi_i \rceil - \lceil \gamma \phi_i \rceil + 1 \right)\frac{L}{r} \qquad (15)$$

As the value of γ decreases faster at large N, the term $W - w_i$ reduces and WFI index (C_i) decreases and hence improves worst case fairness. As far as complexity is concerned, our scheme slightly increases complexity of scheduler as update in γ is required but this update is done once in a Round Robin cycle. Thus it is possible to achieve low latency and improved fairness with little increase in complexity of implementation.

demonstrates the trade-off between latency and data rate. This scheme also improves *worst-case fairness*. From Equation (7), we may write WFI for LL-WRR as

$$C_i = (W - w_i + 1)\frac{L}{r} = \left(\sum_{i=1}^{N} \lceil \gamma\phi_i \rceil - \lceil \gamma\phi_i \rceil + 1\right)\frac{L}{r} \tag{15}$$

As the value of γ decreases faster at large N, the term W-w_i reduces and WFI index (C_i) decreases and hence improves worst case fairness. As far as complexity is concerned, our scheme slightly increases complexity of scheduler as update in γ is required but this update is done once in a Round Robin cycle. Thus it is possible to achieve low latency and improved fairness with little increase in complexity of implementation.

3.1. Architecture of LL-WRR

The overall LL-WRR scheduling architecture is shown in Figure 4 and algorithm is shown in pseudocode 1. Each connection's queue is assigned integer weight w_i whose value is obtained by multiplying its fractional weight ϕ_i with γ and then ceiling the product. Scheduler visits each connection one after the other and removes w_i packets. Once it serves all the queues i.e. when a Round Robin cycle is completed, it determines number of connections that are present and then updates value of γ. Based on updated value, integer weights of all connections are recalculated and process is repeated.

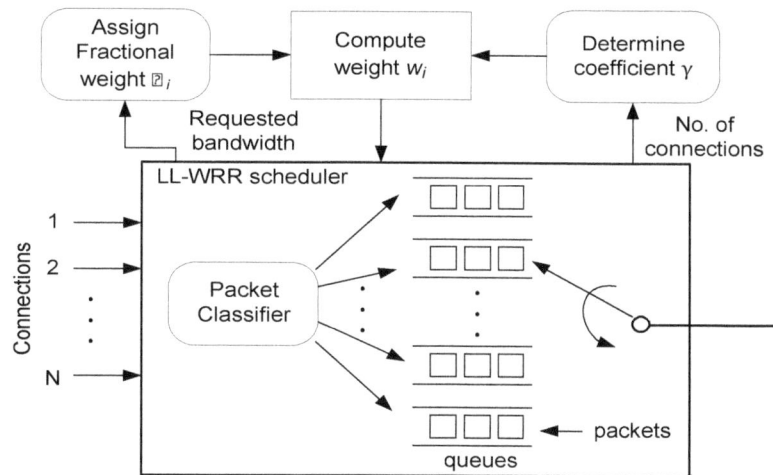

Figure 4. Architecture of LL-WRR scheduling

Pseudocode 1: Algorithmic steps of LL-WRR scheduling

Notations:

N: Number of connections

N_{max} : Maximum number of connections supported

ϕ_i: fractional weight of connection i

w_i : integer weight of connection i

γ : cooefficent

W: sum of w_i or Round Robin length

Q_i: size of queue of ith connection

$PktSize_i$: Size of packet for i^{th} connection

Initialization:

Assign fractional weight ϕ_i to each service queue

N =0;

For all i **do**

$$w_i = \left\lceil \phi_i \left(\gamma_0 \sqrt{1 - \eta [N/N_{max}]^2} \right) \right\rceil$$

End for

Enqueuing:

$Q_i = Q_i + PktSize_i$;

If $(Q_i > Q_{i,max})$ drop the packet;

Dequeuing:

While (True)

If (Nonempty queue exist) {

For all i **do**

if (w_i >0 and Q_i >0)

Transmit packet to output link from queue i;

$w_i = w_i$ -1;

$Q_i = Q_i - PktSize_i$;   /*update queue size*/

End if

End for   /* round is over */

Update N and ϕ_i ;

Compute γ;

For all i do

$w_i = w_i * \gamma$;

End for

End if

End while

---

4. PERFORMANCE EVALUATION

In this section, we evaluate the performance of the proposed LL-WRR scheduling algorithm in the context of IEEE 802.16 MAC layer [1]. More specifically, the effect of number of connections (with varying subscriber stations) on the scheduling algorithms is studied. The LL-WRR algorithm is developed, configured and simulated in NS-2 [15]. Simulation script is written in OTcl for defining wireless network scenario with single 802.16 base station and multiple subscriber stations (SSs) where SSs are *mobile nodes* with average mobility of 5 m/s.

4.1. Simulation Results

The objective of simulation experiments is to evaluate the performance of proposed LL-WRR algorithm and compare it with conventional WRR and deficit round robin (DRR) scheduling algorithms under CBR audio and MPEG4 video traffic. The experiments are conducted with these two types of traffic (flows) generated by each SS. Plots of *data rate, latency* and *fairness* under varying no. of connections (CBR plus video connections) are obtained for LL-WRR.

The data rate plot for each traffic type (Figure 5 and 6) is obtained as a function of number of connections N in network. As seen from plots, data rate performance WRR is slightly better than our scheme and DRR under moderate to high CBR traffic. For MPEG4 traffic, DRR achieves highest data rate whereas data rate of proposed LL-WRR is almost similar to WRR.

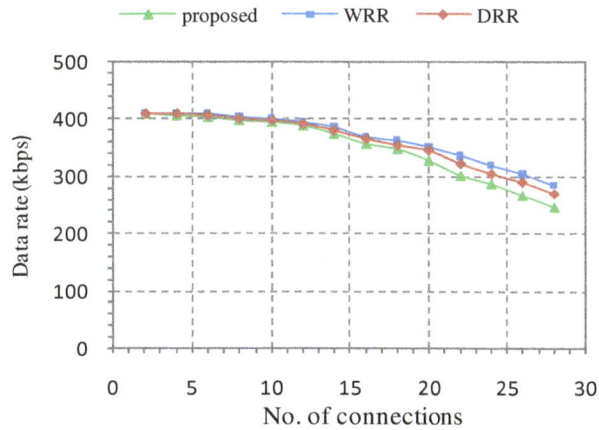

Figure 5. Data rate of CBR audio traffic

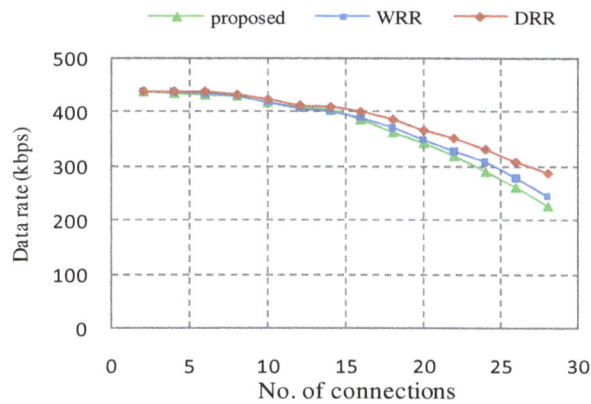

Figure 6. Data rate of MPEG4 video traffic

Proposed scheme exhibits much better latency characteristics than conventional WRR and DRR under CBR traffic and slightly better than DRR under video traffic as evident from plots of Figure 7 and 8. The conventional WRR does not guarantee bounded delay and hence its delay performance is worst than others.

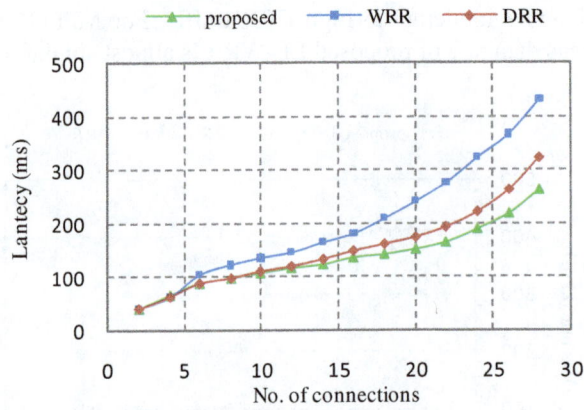

Figure 7. Latency of CBR audio traffic

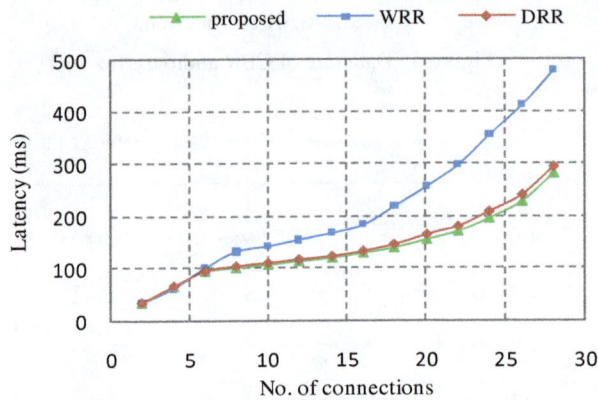

Figure 8. Latency of MPEG4 video traffic

Besides low latency, LL-WRR scheme also offers improvement in fairness as compared to conventional WRR. Figure 9 demonstrates as number connections increases LL-WRR offers much better fairness than conventional WRR. The fairness performance of conventional WRR is inferior to both DRR and LL-WRR. Since DRR takes into account packet length for scheduling, it possesses good fairness property and LL-WRR fairness closely follows DRR.

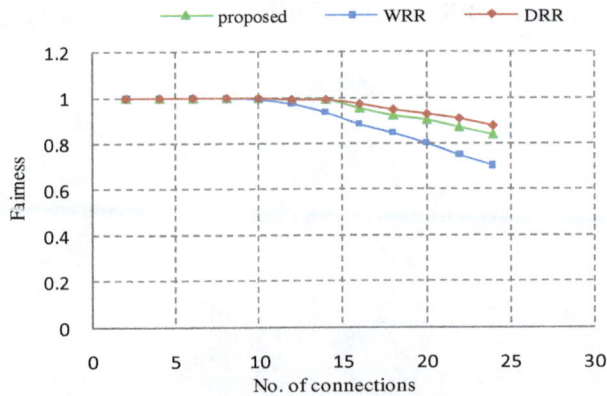

Figure 9. Fairness plot

The mobility of wireless node is an important factor to analyze for understanding its impact on latency. Latency varies with the variation in speed of node; initially increases rapidly up to 9 m/s and the more or less remains steady as shown in Figure 10 and 11. All schemes have this kind of behaviour under both CBR and video traffic. For CBR traffic, both WRR and DRR exhibit similar but larger latency then LL-WRR whereas for video traffic DRR has lower latency than WRR. When speed is increased from 1 m/s to 9 m/s, delay of LL-WRR scheme is increased from 38ms to approximately 150ms under CBR as well as video traffic. Then, it then rises very slowly and tends to remain constant.

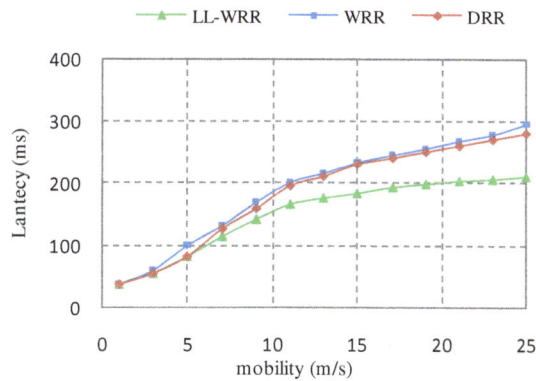

Figure 10. Latency for CBR with mobility (No. of nodes=10)

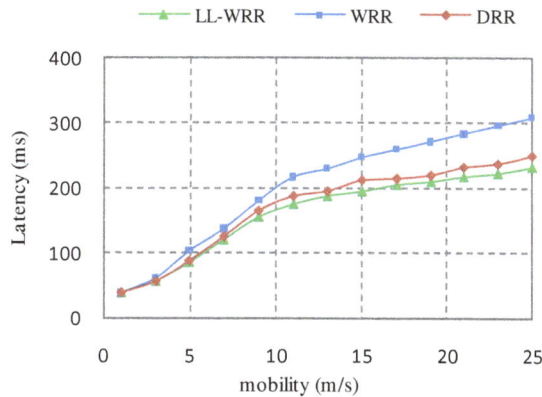

Figure 11. Latency for video with mobility (No. of nodes=10)

5. CONCLUSION

The work of this paper presents simple but efficient scheme named LL-WRR to improve conventional WRR in order to achieve low worst-case latency and improved fairness without much sacrificing rate. However, the computation of coefficient γ introduces additional complexity in proposed scheme but its overall impact will be very small, since it is computed only at the beginning of WRR cycle and not at every packet arrival and departure. The simulation results show that proposed scheme exhibits very low latency than conventional WRR for both CBR audio and MPEG4 video traffic. As compared to DRR, LL-WRR offers less latency for CBR audio traffic. Our scheme also offers better fairness than WRR and remains very close to DRR.

REFERENCES

[1]   IEEE 802.16e-2005, "Local and Metropolitan Networks — Part 16: Air Interface for Fixedand Mobile Broadband Wireless Access Systems, Amendment 2: Physical and Medium Access Control Layers for Combined Fixed and Mobile Operation in Licensed Bands and Corrigendum 1," 2006.

[2]   X. Gao, G. Wu and T. Miki, "End-to-end QoS Provisioning in Mobile Heterogeneous Networks," *IEEE Wireless Communication*, Vol.11, No.3, pp.24-34, June 2004.

[3]   Abhay K. Parekh and R.G.Gallager, "A Generalized Processor Sharing Approach to Flow Control in Integrated Services Networks," *IEEE/ACM Transaction on Networking* Vol.2, No.2, April 1994.

[4]   C.R. Bennett and H. Zhang, "WF2Q: Worst-case fair weighted fair queueing," *IEEE INFOCOM'96,* pp.120-128, March 1996.

[5]   S.J. Golestani, "A self-clocked fair queueing scheme for broadband applications," Proceedings of *IEEE INFOCOM '94*, Vol.2, pp.636-646, June 1994.

[6]   H.M. Alnuweiria and H.Tayyar, "Analysis of virtual-time complexity in weighted fair queuing," Journal of Computer Communications (Elsevier), Vol. 28, No.7, pp.802-810, May, 2005.

[7]   D. Stiliadis and A. Varma, "Efficient fair queueing algorithms for packet-switched networks," *IEEE/ACM Transaction on Networking*, Vol. 6, No.2, pp.175-185, April 1998.

[8]   L.H.Chen, Wu E.H.K, M.I. Hsieh, J T Horng and G H Chen, "Credit-based low latency packet scheduling algorithm for real-time applications," In proc. of IEEE Int. Conf. on Communication, Networks and Satellite (ComNetSat 2012), pp.15-19, July 2012.

[9]   D. Stiliadis and A. Varma, "Latency-rate servers: A general model for analysis of traffic scheduling algorithms," *IEEE/ACM Transaction on Networking*, Vol. 6, No.5, pp.611-624, Oct. 1998.

[10]   H.M. Chaskar and U. Madhaw, "Fair Scheduling with Tunable Latency: A Round Robin Approach," *IEEE /ACM Transaction on networking*, Vol. 11, No.4, pp. 592-601, Aug. 2003.

[11]   W. Mardini and Mai M. Abu Alfool, "Modified WRR Scheduling Algorithm for WiMAX Networks", Journal of Network Protocols and Algorithms, Vol. 3, No.2, pp.24-53 July 2011.

[12]   A. Sayenko, T. Hamalainen, J. Joutsensalo and P. Raatikainen, "Adaptive scheduling using the revenue-based weighted round robin," *Proceedings of IEEE ICON '04*, Vol.2, pp.743-749, Nov. 2004.

[13]   R. Ouni, J. Bhar and K. Torki, "A New Scheduling Protocol Design Based On Deficit Weighted Round Robin For QoS Support In IP Networks," Journal of Circuits Systems and Computers, World Scientific Publishing Company, Vol. 22, No. 3, (2013) 1350012.

[14]   D. Bertsekas and R. Gallager, "Data Networks," 2nd Ed., Prentice Hall, Englewood Cliffs, New Jersey 07632.

[15]   The Network Simulator-NS-2, http://www.isi.edu/nsnam/ns.

18

MACRO WITH PICO CELLS (HETNETS) SYSTEM BEHAVIOUR USING WELL-KNOWN SCHEDULING ALGORITHMS

Haider Al Kim[1], Shouman Barua[2], Pantha Ghosal[2] and Kumbesan Sandrasegaran[2]

[1]Faculty of Engineering and Information Technology, University of Technology Sydney, Australia

ABSTRACT

This paper demonstrates the concept of using Heterogeneous networks *(HetNets) to improve Long Term Evolution (LTE) system by introducing the LTE Advance (LTE-A). The type of HetNets that has been chosen for this study is Macro with Pico cells. Comparing the system performance with and without Pico cells has clearly illustrated using three well-known scheduling algorithms (Proportional Fair PF, Maximum Largest Weighted Delay First MLWDF and Exponential/Proportional Fair EXP/PF). The system is judged based on throughput, Packet Loss Ratio PLR, delay and fairness.. A simulation platform called LTE-Sim has been used to collect the data and produce the paper's outcomes and graphs. The results prove that adding Pico cells enhances the overall system performance. From the simulation outcomes, the overall system performance is as follows: throughput is duplicated or tripled based on the number of users, the PLR is almost quartered, the delay is nearly reduced ten times (PF case) and changed to be a half (MLWDF/EXP cases), and the fairness stays closer to value of 1. It is considered an efficient and cost effective way to increase the throughput, coverage and reduce the latency.*

KEYWORDS

HetNets, LTE &LTE-A, Macro, Pico, Scheduling algorithms & LTE-Sim

1. INTRODUCTION

In the Long Term Evolution so-called LTE, the requirements for larger coverage area, more capacity, and high data rate and low latency have led to search for cost-effective solutions to meet these demands. Hence, the development in the telecommunication networks has adopted different directions to enhance the LTE system taking into account the International Mobile Telecommunications (IMT-2000) standards that have to be satisfied [1]. Network-based technologies such as Multiple Input and Multiple Output MIMO/ advanced MIMO and Transmission/Reception Coordinated Multi-Point CoMP are LTE enhancements that introduce LTE Advance (LTE-A). Other less cost enhancements based on air interfaces are proposed, such as improving spectral efficiency involving using Heterogeneous networks (HetNets). HetNets are small and less power cells within the main macro cells with different access technologies to close up the network to the end users and increase their expectation [16].According to [2], there are two main practical HetNets classes: Macro with Femto and Macro with Pico. Femto and Pico are the small and less power cells. To save the cost, operators use the same carrier frequency in the large and small cells which, on the other hand, proposes interference challenges. Figure 1 gives the main concept of HetNets. To clarify, user in LTE is well-known as a UE.

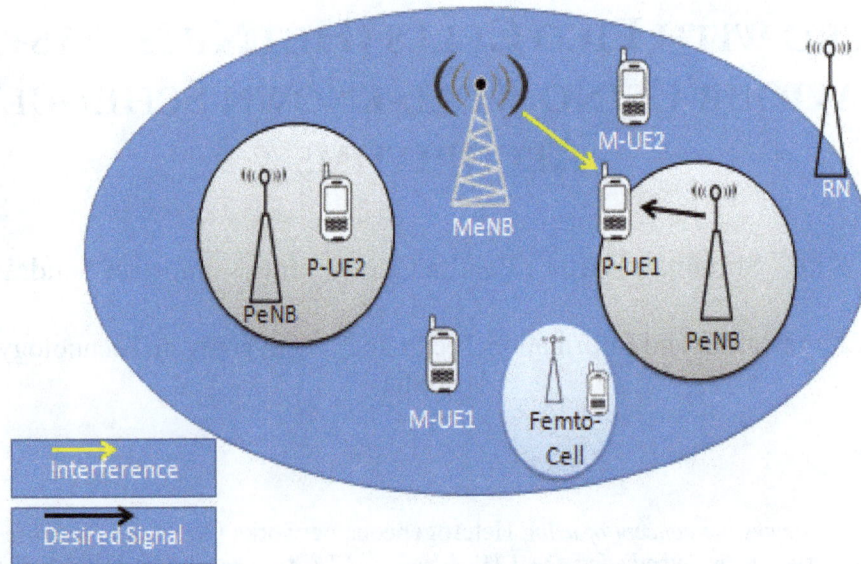

Figure.1 an example of HetNets

In LTE and LTE-A, the element that is responsible for Radio Resources Management (RRM) is enhanced Node Base station (so-called eNB). The eNB does all required management including Packet Scheduling (PS) which is the focus in the paper. PS can guarantee the agreed quality of service demands (QoS) because it is responsible for the best and effective utilizing of the affordable radio resources and in charge of data packets transmission of the users[3].

3rd Generation Partnership Project (3GPP) has left the scheduling algorithms to be vendor specific according to user's requirements and network capability. Therefore, various PS algorithms have been proposed depending on the traffic sorts and provided services. PF, MLWDF and EXP/PF algorithms [4][5][6] are used in this paper to study and compare between the system behaviours in HetNets (single Macro with 2 Pico cells) using these three types of algorithms. Scheduling algorithms ensure that QoS requirements have been met. This can be conducted by prioritizing each link between the eNB and the users, the higher priority connection the first handled in the eNB.

This paper is organized as follows. Section II discusses the downlink system model of LTE. The followed section (III) describes in more details packet scheduling algorithms, while Section IV present simulation environment. Section V shows the outcomes of the simulation. Finally, conclusion is given in Section VI.

2. DOWNLINK SYSTEM MODEL OF LTE

The basic element in the downlink direction of the LTE networks is called Resource Block (RB).Each UE is allocated certain number of resource blocks according to its status, the traffic type and QoS requirements. It could define the RB in both frequency domain and time domain. In the time domain, it comprises single (0.5 ms) time slot involving 7 symbols of OFDMA (orthogonal frequency division multiple access). In the frequency domain, on the other hand, it consists of twelve 15 kHz contiguous subcarriers resulting in 180 kHz as a total RB bandwidth [7].

As aforementioned before, the eNB is responsible for PS and other RRM mechanisms. The bandwidth that is used in this study is 10 MHz considering the inter-cell interference is existed.

The period that eNB performs new packet scheduling operation is the Transmission Time Interval (TTI). TTI is 1 ms that mean the users are allocated 2 contiguous radio resource blocks (2RBs). The scheduling decision in the serving eNB is made based on the uplink direction reports come from the UEs at each transmission time interval. The reports comprise the channel conditions on each RB, such as signal to noise ratio (SNR). The serving eNB uses the SNR value involved in the reports to specify the DL data rate for each served UE in each TTI. For example, how many bits per 2 contiguous RBs [8].

The data rate $dr_i(t)$ for user i at j sub-carrier on RB and at t time can be determined by using equation (1) as proposed in [9].

$$dr_i(t) = A * B * C * D \qquad (1)$$
$$A = nbits_{i,j}(t)/symbol$$
$$B = nsymbols/slot$$
$$C = nslots/TTI$$
$$D = nsc/RBrgg$$

The number of bits per symbol is "A". The number of symbols per slot is "B". While "C" represents how many slots per TTI, "D" clarifies how many sub-carriers per RB. Table 1 summarizes the mapping between SNR values and their associated data rates.

Table 1. Mapping between instantaneous downlink SNR and data rate

Minimum SNR Level (dB)	Modulation and coding	Data Rate (Kbps)
1.7	QPSK (1/2)	168
3.7	QPSK (2/3)	224
4.5	QPSK (3/4)	252
7.2	16 QAM (1/2)	336
9.5	16 QAM (2/3)	448
10.7	16 QAM (3/4)	504
14.8	64 QAM (2/3)	672
16.1	64 QAM (3/4)	756

Upon the packets reach the eNB, they are buffered in eNB in a specific container allocated for each active UE. Moreover, the buffered packets are assigned a time stamp to ensure that they will be scheduled or dropped before the scheduling time interval is expired, and then using First-In-First-Out (FIFO) method they are transmitted to the users in the downlink direction. To explain the scheduling operation, PS manager (is a part of eNB functionalities) at each TTI priorities and classifies the arriving users' packets according to preconfigured scheduling algorithm.

Scheduling decision is made based on different scheduling criteria that have been used in various algorithms. For example channel condition, service type, Head-of-Line (HOL) packet delay, buffer status, and so on so forth. One or more RBs could be allocated to the selected user for transmission with the highest priority. Figure 2 shows the packet scheduler in the downlink direction at eNB.

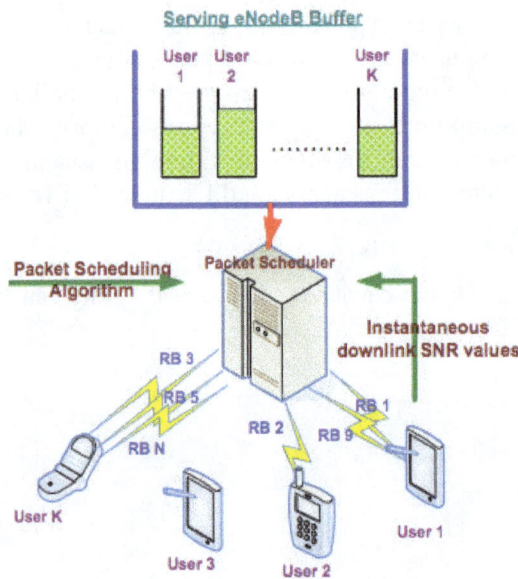

Figure.2 Downlink Packet Scheduler of the 3GPP LTE System [10]

3. PACKET SCHEDULLING ALGORITHMS

The efficient radio resource utilization and ensuring fairness among connected users, as well as satisfying QoS requirements, are the main purposes of using PS algorithms [11].The PS algorithms that have been used in this study are : Proportional Fair (PF) algorithm, Maximum-Largest Weighted Delay First (MLWDF or ML) and the Exponential/Proportional Fair (EXP/PF or EXP) algorithm. It should be noted that these algorithms are used.

3.1. Proportional Fair (PF) Algorithm

For non-real time traffic, the PF was proposed which is used in a Code Division Multiple Access-High Data Rate (CDMA-HDR) system in order to support Non-Real Time (NRT) traffic. In this algorithm, the trade-off between fairness among users and the total system throughput is presented.

This is, before allocating RBs, it considers the conditions of the channel and the past data rate. Any scheduled user in PF algorithm is assigned radio resources if it maximizes the metric k that calculated as the ratio of reachable data rate $r_i(t)$ of user i at time t and average data rate $R_i(t)$ of the same user at the same time interval t:

$$k = \arg\max \frac{r_i(t)}{R_i(t)} \qquad (2)$$

where;

$$R_i(t) = \left(1 - \frac{1}{t_c}\right) * R_i(t-1) + \frac{1}{t_c} * r_i(t-1) \qquad (3)$$

t_c is the window size used to update the past data rates values in which the PF algorithm maximizes the fairness and throughput for any scheduled user. Unless user i is selected for transmission at$(t-1)$, $r_i(t-1) = 0$.

3.2. Maximum Largest Weighted Delay First (MLWDF) Algorithm

If the traffic is a Real Time (RT), the MLWDF is introduced which is used in CDMA-HDR system in order to support RT data users [11].It is more complex algorithms compare with PF and is used in different QoS user's requirements. This is because it takes into account variations of the channel when assigning RBs. Moreover, if a video traffic scenario, it takes into consideration time delay. Any user in MLWDF is granted RBs if it maximizes the equation below:

$$k = \arg\max a_i W_i(t) \frac{r_i(t)}{R_i(t)} \qquad (4)$$

where;

$$a_i = -\frac{(log\delta_i)}{\tau_i} \qquad (5)$$

where $W_i(t)$ is a difference in time between current and arrival times of the packet that known as the Head Of Line (HOL) packet delay of user i at time t.

Similarly to PF equation, while the achievable data rate of user i at time t is $r_i(t)$, the average data rate of the same user at the same time interval t is $R_i(t)$. τ_i and δ_i are the delay threshold for a packet of user i and the maximum HOL packet delay probability of user i respectively. The later is considered to exceed the delay threshold of user i.

3.3. Exponential/Proportional Fair (EXP/PF) Algorithm

Since PF is not designed for multimedia applications (only for NRT traffic), an enhanced PF called EXP/PF algorithm was proposed in the Adaptive Modulation and Coding and Time Division Multiplexing (AMC/TDM) systems. The EXP/PF algorithm is designed for NRT service or RT service (different sorts of services). The k metric is used for both RT nad Non-RT in which RBs are assigned to users based on k.

$$k = \arg\max \begin{cases} \exp\left(\frac{a_i W_i(t) - a\overline{W(t)}}{1+\sqrt{aW(t)}}\right) \frac{r_i(t)}{R_i(t)} & i\epsilon RT \\ \frac{w(t)}{M(t)} \frac{r_i(t)}{R_i(t)} & i\epsilon NRT \end{cases} \qquad (6)$$

where,

$$a\overline{w(t)} = \frac{1}{N_{RT}}\Sigma_{i\epsilon RT} a_i W_i(t) \qquad (7)$$

$$w(t) = \begin{cases} w(t-1) - \varepsilon & W_{max} > \tau_{max} \\ w(t-1) + \frac{\varepsilon}{k} W_{max} < \tau_{max} \end{cases} \qquad (8)$$

where the average number of packets at the buffer of the eNB at time t is represented by $M(t)$, k and ε in equation (8) are constants, $W_i(t)$ is explained in MLWDF, W_{max} is the HOL packets delay of RT service and τ_{max} is the maximum delay of RT service users. The EXP/PF differentiates between RT and NRT by prioritizing RT traffic users over the NRT traffic users if their HOL values are reaching the delay threshold.

4. SIMULATION ENVIRONMENT

LTE-Sim simulator is used in this paper to do the entire analysis and study [12]. The most recent version of LTE-Sim (version 5) has not involved yet any code regarding the HetNets type (Macro with Pico cells). The developed code used in this paper could be considered as an enhancement of the released LTE-Sim versions. However, LTE-Sim has a detailed code (or what authors are named it: scenario) which can be used to simulate and examine HetNets type (Macro with Femto). Our paper is based on a scenario of a single Macro cell with 2 small Pico cells that are reduced their powers. More Picos can be added to the system, and enhanced system behaviour will be presented. However, according to [2], while the number of Pico cells is increased, more inter-cell interference is experienced since the same carrier frequency is used in each cell (Macro and Picos).

Figure 3 shows the entire system that is used in this paper: Macro cell of 1 km and 2 Pico cells of 0.1 km located on the Macro edge. This design is chosen to analog a real system aimed to cover larger area and more users, especially the users in the cell edge where they suffer from lack of connectivity with Macro cell. The inter-cell interference is modeled. Video and VoIP traffic are used to represent user's data. Each user has 50 % Video traffic and 50% VoIP flows.

Handover is activated. Each cell starts a certain number of users. Non-uniform user distribution within the cells is applied and 3km/h constant speed is utilized as the mobility user speed. In addition, the 3GPP urban Macro cell propagation loss model has been implemented including path-loss, penetration loss, multi-path loss and shadow fading which are summarized below [13]:

- Pathloss: $128.1 + 37.6 \log_{10}(d)$, d refers the distance between the eNB and the user in kilometers.
- Penetration loss: 10 dB
- Multipath loss: using one of the well-known methods called *Jakes model*
- Shadow fading loss (recently it could be used as a gain in LTE-A): *log-normal distribution*
  - *Mean value of 0 dB.*
  - *Standard deviation of 10 dB.*

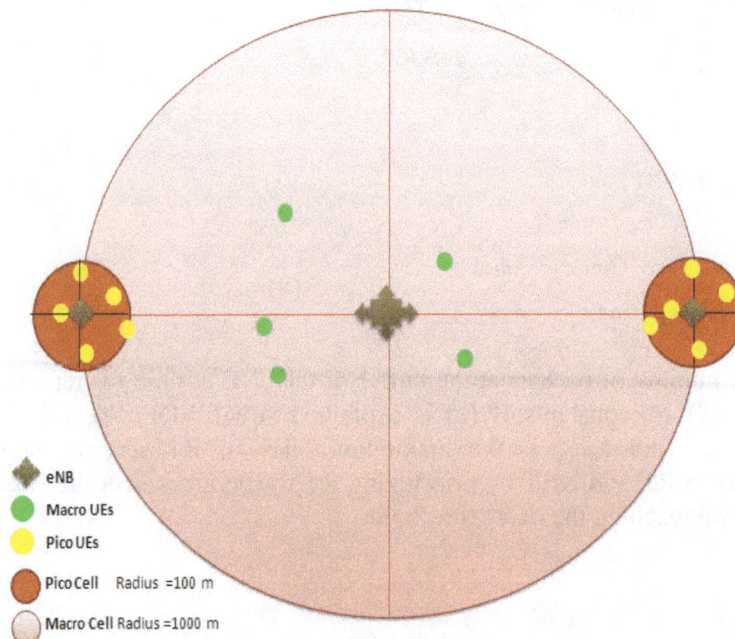

Figure.3 Applied HetNets (Macro with 2 Picos)

Packets throughput (see equation 9), Packet Loss Ratio (PLR) as shown in equation 10, packet delay (latency) and fairness index (equation 11) are the concepts used in the aforementioned algorithms to evaluate the system performance. Jain's method is applied to implement fairness among users [14]. According to [1], fairness should reach the value of 1 to be considered as a fair algorithm that sharing the resources suitably among users. It can be calculated as value 1 minus the value of the difference between the maximum and minimum size of transmitted packets of the most and least scheduled users. Equation (11) calculates the fairness value.

$$throughput = \frac{1}{T} \sum_{i=1}^{K} \sum_{t=1}^{T} ptransmit_i(t) \tag{9}$$

$$PLR = \frac{\sum_{i=1}^{K} \sum_{t=1}^{T} pdiscard_i(t)}{\sum_{i=1}^{K} \sum_{t=1}^{T} psize_i(t)} \tag{10}$$

$$fairness = 1 - \frac{ptotaltransmit_{max} - ptotaltransmit_{min}}{\sum_{i=1}^{K} \sum_{t=1}^{T} psize_i(t)} \tag{11}$$

Obviously, while $ptransmit_i(t)$ is the size of transmitted packets, $pdiscard_i(t)$ is the size discarded or lost packets during the connection. $psize_i$ is the summation of all arrived packets that are buffered into serving eNB [1].

The aforementioned total size of transmitted packets of the best served UE and the worse served UE are represented in equation (11) as $ptotaltransmit_{max}$ and $ptotaltransmit_{min}$.

Table 2 shows the entire system simulation parameters [1].

Table 2. LTE system simulation parameters

Parameters	
Simulation time	30 s
Flow duration	20 s
Slot duration	0.5 ms
TTI	1 ms
Number of OFDM symbols/slot	7
Macro cell radius	1 km
Macro eNB Power	49 dBm
Pico cell radius	0.1 km
Pico eNB Power	30 dBm
User speed	3 km/h
VoIP bit rate	8.4 kbps
Video bit rate	242 kbps
Frame structure type	FDD
Bandwidth	10 MHz
Number of RBs	50
Number of subcarriers	600
Number of subcarriers/RB	12
Subcarrier spacing	15 KHz

In order to get better results and to confirm the outcomes, five simulations have been conducted for each algorithm (PF, MLWDF and EXP) in each point of users (10, 20, 30, 40, 50, 60, 70 and

80). This yields 120 simulations outcomes. The average values have been taken to draw the simulation graphs at each point of users.

5. SIMULATION RESULTS

The average overall system throughput is shown in figure 4. Comparing the throughput for "single Macro cell" for the same simulation parameters as shown in figure 5, the pico cells in the scenario "Macro with 2 Picos" boost the throughput by adding gain that shown as an overall system throughput increment for the same number of users.   For instance, at 40 users using MLWDF, the throughput is 25 Mbps for the scenario with 2 Picos while the Macro scenario is only 9.3 Mbps. This is almost a duple value. Further points show duple and triple throughput values in the scenario of 2 Picos. However, the gain will reach a saturation level where no more gain could be shown due to the fact of limited radio resources availability while more users are added to the system. Although MLWDF and EXP have almost similar behaviour in both scenarios, a higher throughput is shown in the 2 Pico case using both algorithms. It could note that PF algorithm as shown figure 5 behaves better than the scenario of single Macro cell. PF is developed for NRT traffic, but the simulation is for Video flows (RT traffic); hence, the other simulated algorithms outperform PF.

PLR shown in the figure 6 according to [15] is the packet loss ratio for a single Macro cell. While the system is charged with more than 20 users, the PLR is increased for all experienced algorithms taking into consideration that the PF is the worst case with the video traffic. Adding two Picos to the previous system to create "Macro with 2 Picos" scenario enhances the PLR while maintaining similar system behavior for all algorithms. Approximately, the PLR in Macro with 2 Picos case is reduced to be a quarter of PLR value of single Macro cell scenario. For example, at 70 users, MLWDF has 0.1 PLR value while for the same number of users MLWDF has 0.5 PLR value in the single Macro scenario. Comparing between scheduling schemes, the worst case is the PF algorithm in both cases. Figure 7 illustrates PLR for Macro with 2 Picos.

According to [15] and as shown in figure 8 , the delay in single Macro cell scenario is close to be constant for PF, MLWDF and EXP/PF with value less than 5 ms while it suffers from rapid increasing after 40 users for PF algorithm. If two Pico cells are added to the aforementioned system, a similar performance is shown, but the delay value is decreased. In addition, the threshold of PF is shifted at 60 users instead of 40 users in the single Macro case. To compare MLWDF and EXP/PF in both scenarios, a certain point in figures 8 and 9 could be explained. For example at 60 users, in a single Macro cell the delay value is 50 ms while in the Macro with 2 Picos the value is 20 ms. As a consequence, for MLWDF and EXP/PF, the delay value with two Picos is approximately half the delay value without Pico cells. One of the purposes of HetNets is to enhance the latency, and this is shown in a practical simulation illustrated in figure 9. However, the delay shows lower values (nearly 10 times lower) in the scenario of single cell with 2 Picos using PF scheme.

When the number of users increases in single Macro cell more than 30, the fairness index of all simulated algorithms is deviated down of the value "1". At 40 users, PF shows further deviation close to value 0.8 compare with other algorithms which they are around 0.9. The fairness index behaves similarly in the scenario of Macro with 2 Picos as shown in figure 11. However, the PF shows a minor different in which at 50 users it starts to decline to get the value 0.8.

5.1. Throughput

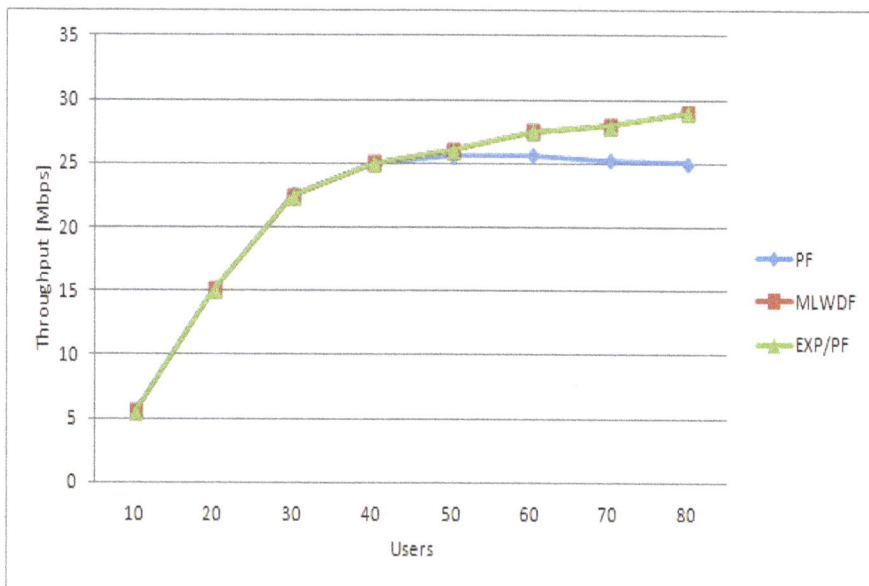

Figure.4 Average System Throughput (Macro with 2 Picos)

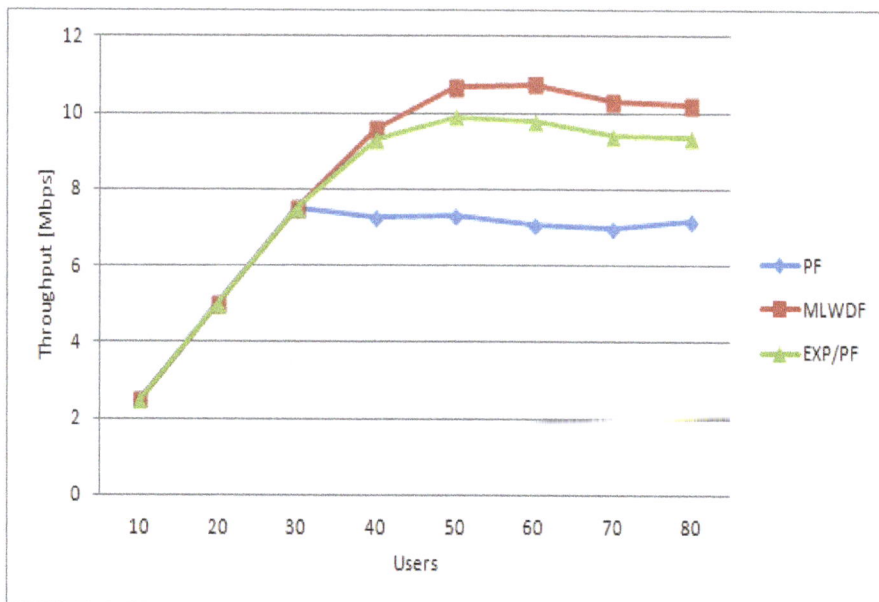

Figure.5 Average System Throughput (single Macro cell)

5.2. Packet Loss Ratio (PLR)

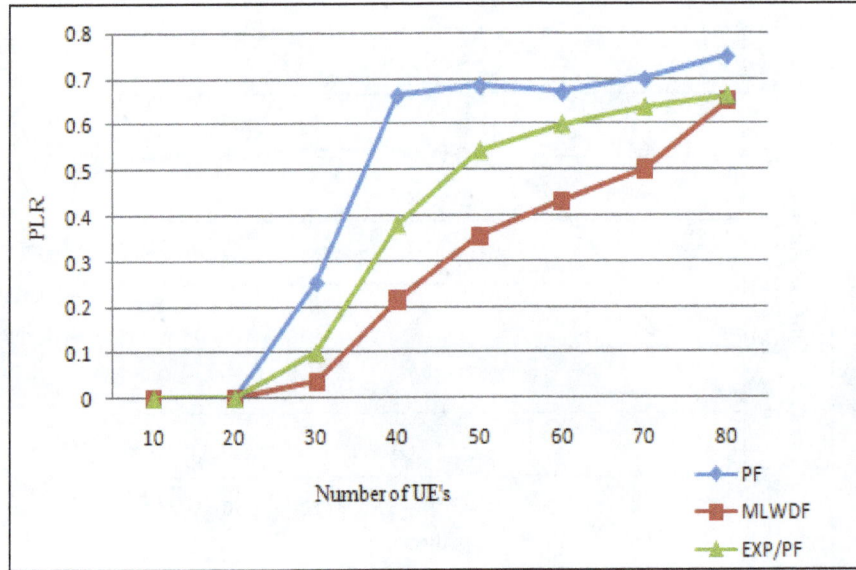

Figure.6 PLR of Video Flows (single Macro cell) [15]

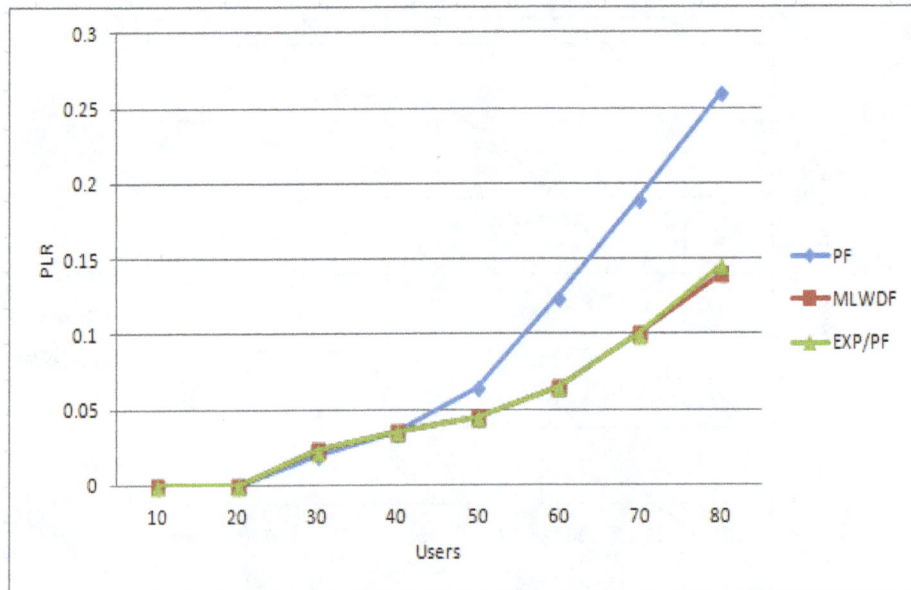

Figure.7 PLR of Video Flows (Macro with 2 Picos)

5.3. Delay

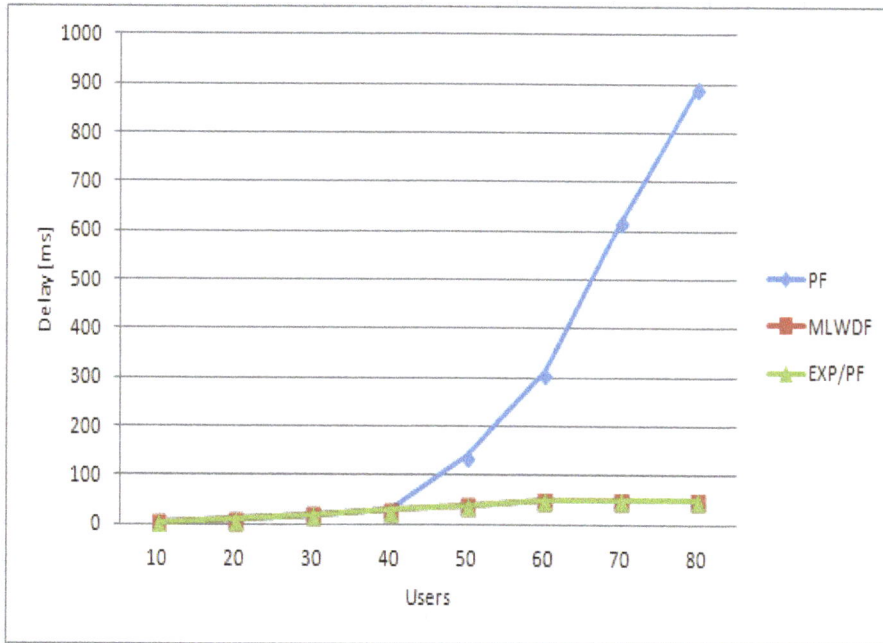

Figure.8 Packet Delay of Video Flows (single Macro cell)

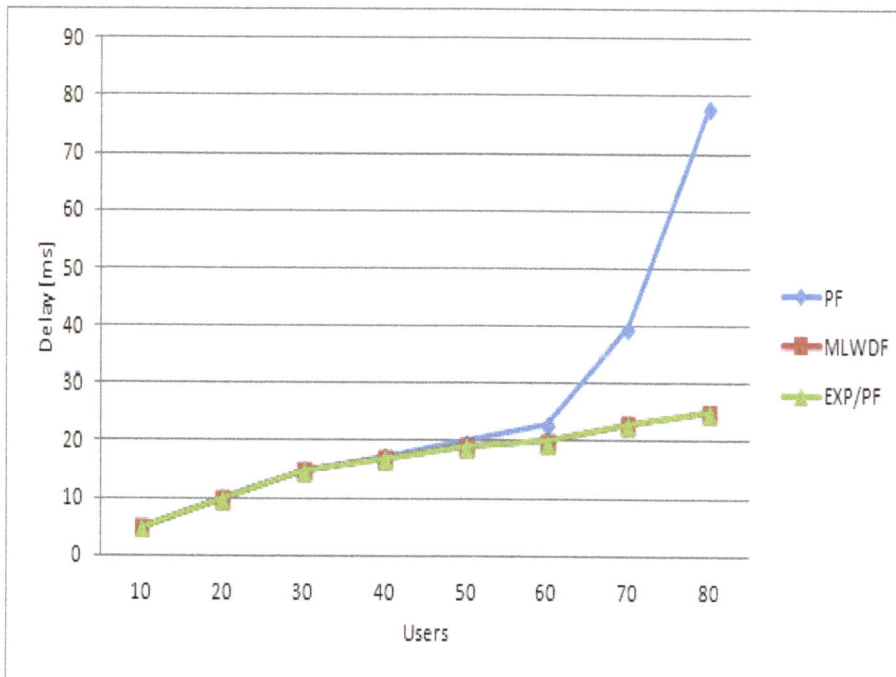

Figure.9 Packet Delay of Video Flows (Macro with 2 Picos)

5.4. Fairness Index

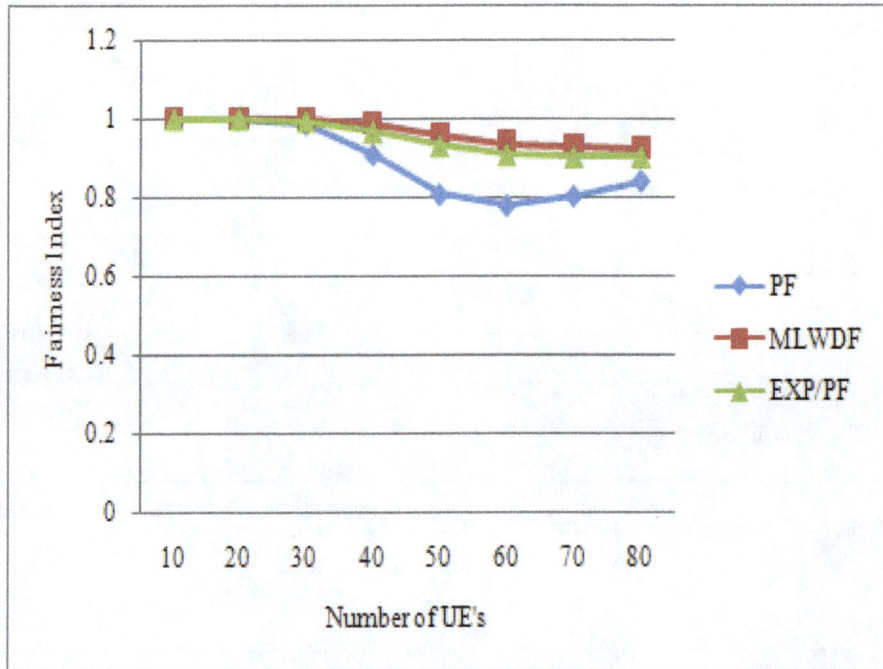

Figure.10 Fairness Index of Video Flows [15]

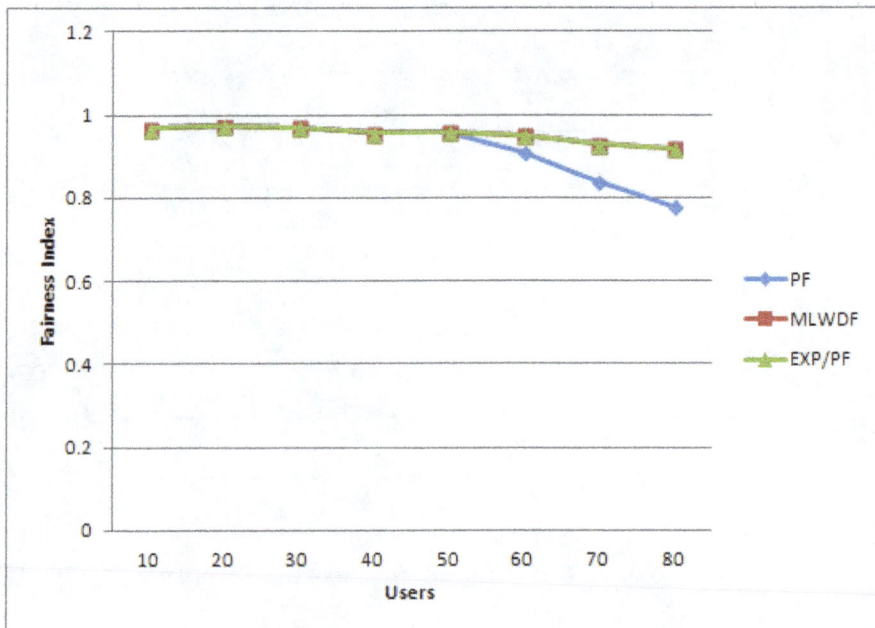

Figure.11 Fairness Index of Video Flows Macro with 2 Picos

6. CONCLUSION

This paper investigates scheduling algorithms that are developed to enhance the LTE network performance by sharing radio resources fairly among users utilizing all available resources. These algorithms depend on traffic class and number of users, hence; different outcomes are presented for each algorithm. To further boost the overall system performance, this study uses heterogeneous networks concept by adding small cells (2 Pico cells). This enhancement is experienced through a throughput, PLR, delay and fairness. In the throughput the system gains more data rate while in PLR the system suffers less packet loss values. Moreover, delay is decreased and fairness stays similar. Approximately from the simulation outcomes, the overall system performance is as follows: throughput is duplicated or nearly tripled relaying on the number of users, the PLR is almost quartered, the delay is reduced 10 times (PF case) and changed to be a half value (MLWDF/EXP cases), and the fairness stays closer to value of 1. As number of small cells increases, the system is expected to be more enhanced till a saturation state is reached. The reason behind that is the inter-cell interference will limit the performance since the same carrier frequency is used in all system's cells. Focusing on macro with 2 Pico cells scenario, MLWDF shows the best performance for video flows followed by EXP/PF. Further enhancement can be applied in future papers such as almost blank subframes (ABS), enhanced inter-cell interference cancelation (eICIC) and cell range extension CRE concepts.

REFERENCES

[1]   H. A. M. Ramli, R. Basukala, K. Sandrasegaran, and R. Patachaianand, "Performance of well known packet scheduling algorithms in the downlink 3GPP LTE system," in Communications (MICC), 2009 IEEE 9th Malaysia International Conference on, 2009, pp. 815-820.

[2]   Seung June Yi, S.C., Young Dae Lee,   Sung Jun Park, Sung Hoon Jung 2012, Radio Protocols for LTE and LTE-Advanced.

[3]   [1]   B. Liu, H. Tian, and L. Xu, "An efficient downlink packet scheduling algorithm for real time traffics in LTE systems," in Consumer Communications and Networking Conference (CCNC), 2013 IEEE, 2013, pp. 364-369.

[4]   A. Jalali, R. Padovani, and R. Pankaj, "Data throughput of CDMA-HDR a high efficiency-high data rate personal communication wireless system," in Vehicular Technology Conference Proceedings, 2000. VTC 2000-Spring Tokyo. 2000 IEEE 51st, 2000, pp. 1854-1858.

[5]   M. Andrews, K. Kumaran, K. Ramanan, A. Stolyar, P. Whiting, and R. Vijayakumar, "Providing quality of service over a shared wireless link," Communications Magazine, IEEE, vol. 39, pp. 150-154, 2001.

[6]   J.-H. Rhee, J. M. Holtzman, and D. K. Kim, "Performance analysis of the adaptive EXP/PF channel scheduler in an AMC/TDM system," Communications Letters, IEEE, vol. 8, pp. 497-499, 2004.

[7]   J. Zyren and W. McCoy, "Overview of the 3GPP long term evolution physical layer," Freescale Semiconductor, Inc., white paper, 2007.

[8]   B. Riyaj, M. R. H. Adibah, and S. Kumbesan, "Performance analysis of EXP/PF and M-LWDF in downlink 3GPP LTE system," 2009.

[9]   X. Qiu and K. Chawla, "On the performance of adaptive modulation in cellular systems," Communications, IEEE Transactions on, vol. 47, pp. 884-895, 1999.

[10]   S. C. Nguyen, K. Sandrasegaran, and F. M. J. Madani, "Modeling and simulation of packet scheduling in the downlink LTE-advanced," in Communications (APCC), 2011 17th Asia-Pacific Conference on, 2011, pp. 53-57.

[11]   A. Alfayly, I.-H. Mkwawa, L. Sun, and E. Ifeachor, "QoE-based performance evaluation of scheduling algorithms over LTE," in Globecom Workshops (GC Wkshps), 2012 IEEE, 2012, pp. 1362-1366.

[12]   G. Piro, L. A. Grieco, G. Boggia, F. Capozzi, and P. Camarda, "Simulating LTE cellular systems: an open-source framework," Vehicular Technology, IEEE Transactions on, vol. 60, pp. 498-513, 2011.

[13]   M. Iturralde, T. Ali Yahiya, A. Wei, and A. Beylot, "Resource allocation using shapley value in LTE networks," in Personal Indoor and Mobile Radio Communications (PIMRC), 2011 IEEE 22nd International Symposium on, 2011, pp. 31-35.

[14]   R. Jain, D.-M. Chiu, and W. R. Hawe, A quantitative measure of fairness and discrimination for resource allocation in shared computer system: Eastern Research Laboratory, Digital Equipment Corporation, 1984.

[15]   AL-Jaradat, Huthaifa 2013, 'On the Performance of PF, MLWDF and EXP/PF algorithms in LTE'.

[16]   Holma H, Toskala A 2012, "LTE-Advanced 3GPP Solution for IMT-Advanced".

Permissions

The contributors of this book come from diverse backgrounds, making this book a truly international effort. This book will bring forth new frontiers with its revolutionizing research information and detailed analysis of the nascent developments around the world.

We would like to thank all the contributing authors for lending their expertise to make the book truly unique. They have played a crucial role in the development of this book. Without their invaluable contributions this book wouldn't have been possible. They have made vital efforts to compile up to date information on the varied aspects of this subject to make this book a valuable addition to the collection of many professionals and students.

This book was conceptualized with the vision of imparting up-to-date information and advanced data in this field. To ensure the same, a matchless editorial board was set up. Every individual on the board went through rigorous rounds of assessment to prove their worth. After which they invested a large part of their time researching and compiling the most relevant data for our readers.

The editorial board has been involved in producing this book since its inception. They have spent rigorous hours researching and exploring the diverse topics which have resulted in the successful publishing of this book. They have passed on their knowledge of decades through this book. To expedite this challenging task, the publisher supported the team at every step. A small team of assistant editors was also appointed to further simplify the editing procedure and attain best results for the readers.

Apart from the editorial board, the designing team has also invested a significant amount of their time in understanding the subject and creating the most relevant covers. They scrutinized every image to scout for the most suitable representation of the subject and create an appropriate cover for the book.

The publishing team has been an ardent support to the editorial, designing and production team. Their endless efforts to recruit the best for this project, has resulted in the accomplishment of this book. They are a veteran in the field of academics and their pool of knowledge is as vast as their experience in printing. Their expertise and guidance has proved useful at every step. Their uncompromising quality standards have made this book an exceptional effort. Their encouragement from time to time has been an inspiration for everyone.

The publisher and the editorial board hope that this book will prove to be a valuable piece of knowledge for researchers, students, practitioners and scholars across the globe.

List of Contributors

Lahby Mohamed
Computer Science Department, LIM Lab Faculty of Sciences and Technology of Mohammedia, B.P. 146 Mohammedia, Morocco

Cherkaoui Leghris
Computer Science Department, LIM Lab Faculty of Sciences and Technology of Mohammedia, B.P. 146 Mohammedia, Morocco

Adib Abdellah
Computer Science Department, LIM Lab Faculty of Sciences and Technology of Mohammedia, B.P. 146 Mohammedia, Morocco

L.Nithyanandan
Department of Electronics and Communication Engineering, Pondicherry Engineering College, Puducherry, India

V.Bharathi
Department of Electronics and Communication Engineering, Pondicherry Engineering College, Puducherry, India

P.Prabhavathi
Department of Electronics and Communication Engineering, Pondicherry Engineering College, Puducherry, India

Pawan Sharma
Department of Electronics and Communication Engineering, Bhagwan Parshuram Institute of Technology, Rohini, New Delhi, India

Seema Verma
Department of Electronics and Communication Engineering, Banasthali University, Rajasthan, India

Cyrine Lahsini
Signal and Communication Department, Telecom Bretagne, France

Sonia Zaibi
Syscoms Laboratory, National Engineering School of Tunis, Tunisia

Ramesh pyndiah
Signal and Communication Department, Telecom Bretagne, France

Ammar Bouallegue
Syscoms Laboratory, National Engineering School of Tunis, Tunisia

B. Siva Kumar Reddy
Department of Electronics and Communication Engineering National Institute of Technology Warangal, Andhra Pradesh-506004, India

B. Lakshmi
Department of Electronics and Communication Engineering National Institute of Technology Warangal, Andhra Pradesh-506004, India

Hayder J. Albattat
Department of Electrical Engineering, Basarah University, Basrah, Iraq

Haider M. AlSabbagh
Department of Electrical Engineering, Basarah University, Basrah, Iraq

S. A. Alseyab
Department of Electrical Engineering, Basarah University, Basrah, Iraq

Radhia GHARSALLAH
National Engineering School of Tunis, Innov'Com Laboratory, Higher School of Communications, Tunisia

Ridha BOUALLEGUE
Innov'Com Laboratory, Higher School of Communications, Tunisia

Tony Tsang
Hong Kong Polytechnic University, Hung Hom, Hong Kong

Bassam A.alqaralleh
Al-Hussein Bin Talal University

Khaled Almi'ani
Al-Hussein Bin Talal University

Prof.Shubhangi Mahamuni
Assist. Prof, Dept.of E&TC, MAE, Alandi (D), Pune,MS

Dr.Vivekanand Mishra
Associate Professor, Dept.of Electronics, SVNIT, Surat,Gujrat

Dr.Vijay M.Wadhai
Principal, MITCOE, Kothrud,Pune, MS

AKM Arifuzzaman
Department of Electrical and Electronic Engineering American International University-Bangladesh, Banani, Dhaka, Bangladesh

Rumana Islam
Department of Electrical and Electronic Engineering American International University-Bangladesh, Banani, Dhaka, Bangladesh

Mohammed Tarique
Department of Electrical Engineering Ajman University of Science and Technology, Fujairah, UAE

Mussab Saleh Hassan
Department of Electrical Engineering Ajman University of Science and Technology, Fujairah, UAE

Indrani Das
School of Computer and Systems Sciences Jawaharlal Nehru University, New Delhi, India

D.K Lobiyal
School of Computer and Systems Sciences Jawaharlal Nehru University, New Delhi, India

C.P Katti
School of Computer and Systems Sciences Jawaharlal Nehru University, New Delhi, India

Alireza Ghodratabadi
EE institute, Malekashtar University of Technology, Tehran, Iran

Hashem Moradmand Ziyabar
IRIB, Tehran, Iran

Jehad Hamamra
Electrical Engineering Department Texas A&M University at Qatar

Hassan El-Sallabi
Electrical Engineering Department Texas A&M University at Qatar

Khalid Qaraqe
Electrical Engineering Department Texas A&M University at Qatar

Mihir N. Mohanty
ITER, Siksha 'O' Anusandhan University, Bhubaneswar, Odisha, 751030, India

Monalisa Bhol
ITER, Siksha 'O' Anusandhan University, Bhubaneswar, Odisha, 751030, India

Laxmi Prasad Mishra
Seemanta Engineering College, Jharpokharia, Mayurbhanj, Odisha, India

Sanjat Kumar Mishra
Seemanta Engineering College, Jharpokharia, Mayurbhanj, Odisha, India

Moses Ekpenyong
Department of Computer Science, University of Uyo, Uyo Academic Visitor, School of Informatics, University of Edinburgh, Edinburgh

Joseph Isabona
Department of Basic Sciences, Benson Idahosa University, Benin City, Nigeria

Imeh Umoren
Department of Computer Science, Akwa Ibom State University, Mkpat Enin, Nigeria

Zuber Patel
Department of Electronics Engg., National Institute of Technology, Surat, India

Upena Dalal
Department of Electronics Engg., National Institute of Technology, Surat, India

Haider Al Kim
Faculty of Engineering and Information Technology, University of Technology Sydney, Australia

Shouman Barua
Faculty of Engineering and Information Technology, University of Technology Sydney, Australia

Pantha Ghosal
Faculty of Engineering and Information Technology, University of Technology Sydney, Australia

Kumbesan Sandrasegaran
Faculty of Engineering and Information Technology, University of Technology Sydney, Australia